# GENETIC NEUROBIOLOGY

# GENETIC NEUROBIOLOGY

Jeffrey C. Hall, Ralph J. Greenspan,
and William A. Harris

This text is based in part on ideas developed at a Neurosciences
Research Program Work Session chaired by Seymour Benzer,
Richard L. Sidman, and Jeffrey C. Hall

The MIT Press
Cambridge, Massachusetts
London, England

## Acknowledgment of Sponsorship and Support

The Neurosciences Research Program, a research center of the Massachusetts Institute of Technology, is an interdisciplinary, interuniversity organization with the primary goal of facilitating the investigation of how the nervous system mediates behavior including the mental processes of man. To this end, the NRP, as one of its activities, conducts scientific meetings to explore crucial problems in the neurosciences and publishes reports of these Work Sessions. NRP is supported in part through Massachusetts Institute of Technology by National Institute of Mental Health Grant No. MH23132, National Institute of Neurological and Communicative Disorders and Stroke Grant NS15690, The March of Dimes Birth Defects Foundation, and Max-Planck-Gesellschaft, and through the Neurosciences Research Foundation by The Camille and Henry Dreyfus Foundation, Inc., The Beverly and Harvey Karp Foundation, van Ameringen Foundation, Inc., The G. Unger Vetlesen Foundation, and Vollmer Foundation, Inc.

This book was set in Compugraphic English Times by Wilkins and Associates, and printed and bound by The Murray Printing Company in the United States of America.

**Library of Congress Cataloging in Publication Data**

Hall, Jeffrey C.
  Genetic neurobiology.

  "Based in part on ideas developed at a Neurosciences Research Program work session."
  Bibliography: p.
  Includes indexes.
  1. Neurogenetics.  2. Behavior genetics.
I. Greenspan, Ralph J.  II. Harris, William A. (William Anthony)  III. Neurosciences Research Program.  IV. Title. [DNLM: 1. Genetics, Behavioral. 2. Nervous system—Physiology.  QH 457 H177g]
QP356.H34       591.51'015751       81-18598
ISBN 0-262-08111-3                  AACR2

# CONTENTS

QP356
·H34

# PREFACE

The idea for this book originated at a Work Session on neurogenetics held at the Neurosciences Research Program under the chairmanship of Seymour Benzer, Richard L. Sidman, and Jeffrey C. Hall, 22–24 October 1978. Many then current facts from the disparate areas of behavioral genetics and genetic neurobiology were consolidated at this meeting. Just as important were several of the thoughts bandied about during the sessions, such as the relation of genes to the development and function of excitable cells and tissues, and how experiments on genetic variants often lead to unique kinds of information.

Proceeding from these ideas, and the bare outline of a book suggested by them, we have considerably expanded on the subject of neurogenetics. We have also tried to integrate the information obtained from a variety of organisms into each separate area of behavioral and neural biology that has been approached genetically. In doing so, we modestly suggest that we have at least departed from the mode of presentation that depends solely on the order of speakers at a scientific conference.

We have also attempted to keep the book relatively up to date, comprehensible, and accurate. In this regard, we appreciate the help we have received from Bob Horvitz, John Palka, Ching Kung, Ursula Drager, Ron Konopka, Chip Quinn, and Seymour Benzer. With respect to the work on the manuscript, in its general outlook and many specific details, we are extremely grateful for the vigorous and thoughtful aid provided by Yvonne Homsy, George Adelman, and Stephen Dennis.

Jeffrey C. Hall
Ralph J. Greenspan
William A. Harris

California
1981

# PARTICIPANTS
## AT THE NRP WORK
## SESSION ON NEUROGENETICS

Julius Adler
Department of Biochemistry
University of Wisconsin,
 Madison

Seymor Benzer
Division of Biology
California Institute of
 Technology

Barton Childs
Department of Pediatrics
The Johns Hopkins University

Stephen G. Dennis*
Department of Anesthesiology
University of Washington

Ursula C. Dräger
Department of Neurobiology
Harvard Medical School

Ralph J. Greenspan
Department of Physiology
University of California Medical
 Center

Jeffrey C. Hall
Department of Biology
Brandeis University

William A. Harris
Department of Biology
University of California,
San Diego

H. Robert Horvitz
Department of Biology
Massachusetts Institute of
 Technology

R. Kevin Hunt
Department of Biophysics
The Johns Hopkins University

Lily Yeh Jan
Department of Physiology
University of California Medical
 Center

Yuh-Nung Jan
Department of Physiology
University of California Medical
 Center

Ronald J. Konopka
Division of Biology
California Institute of
 Technology

Ching Kung
Laboratory of Molecular Biology
University of Wisconsin

Richard J. Mullen
Department of Anatomy
University of Utah Medical
 School

William L. Pak
Department of Biological
 Sciences
Purdue University

*NRP Staff Coordinator

John Palka
Department of Zoology
University of Washington

William G. Quinn
Department of Biology
Princeton University

Francis O. Schmitt
Neurosciences Research Program

Carla J. Shatz
Department of Neurobiology
Stanford University Medical
   School

Richard L. Sidman
Department of Neurosciences
Children's Hospital Medical
   Center

Frederic G. Worden
Neurosciences Research Program

# GENETIC NEUROBIOLOGY

The pursuit of any one scientific field under an injunction against trespassing into another is neither rational nor productive, especially if both have common objects.

Paul Weiss

Individual genes are dumb, but the whole genome is smart.

Seymour Benzer

My genes are blue. My nucleic acids are not my own. It is totally amazing that I know anything, especially that "my" DNA created me practically de novo, from an amorphous egg to a contemplating fool with a tangle of wires for a brain.

*Notes from the Oblivion,* G. Guillaume

# 1
# INTRODUCTION

The neurogenetic approach to understanding the development, structure, and function of excitable cells and systems is a relatively new field of biology. During the last two decades, genetic variants affecting neurobiological and behavioral phenomena began to be investigated in a systematic fashion. Several biologists began independently to isolate behavioral mutants in bacteria, protozoa, roundworms, insects, and other organisms. These mutants were generally induced with chemical mutagens and selected in behavioral assays found in the older literature or newly invented. The general rationale was that, if one could generate large numbers of new behavioral variants, one could create many new ways to investigate neural development and function. Neurogenetics, as the term will be used in this book, is the arbitrary joining together of the principles and results that have been amassed during this period.

From its early beginning as behavioral genetics, neurogenetics moved into neurobiological areas rather quickly because of successful attempts to determine, at least preliminarily, the abnormalities of excitable cells and tissues that underlie the behavioral deficits in many of the new mutant strains. Thus, many of the behavioral mutants have, in a sense, become neurophysiological, neuroanatomical, or neurochemical mutants. Neurobiological information on organisms expressing these altered genotypes has begun to contribute to our knowledge of how nervous systems assemble themselves and how their component cells function. Surprisingly, sophisticated genetic principles and techniques involving these mutants have also been essential in studies aimed at understanding neural defects.

The study of the genetic aspects of abnormal behavior and nervous systems was undertaken many decades before the beginnings of "modern" neurogenetics. Biologists have long recognized behavioral variants with genetic etiologies in organisms as diverse as fruit flies and humans. Studies of defective nervous systems, especially in several behavioral mutants of mouse, were undertaken well in advance of the recent searches for large numbers of new mutants. Significant mutants of neurobiological interest have also been discovered, over the years, on the basis of nonbehavioral criteria, such as defects in anatomy or embryology. Several of these genetic variants, as well as the early behavioral mutants that have been recognized sporadically, will be discussed in this book in their appropriate contexts. Yet in spite of the lengthy history of behavioral and neurological genetics, the new mutants are extremely important because they have done more than simply expand the existing collection of experimental tools for disrupting nerve cells and tissues. The recent success not only in isolating new mutants but also in characterizing them neurobiologically has brought to light in an organized fashion the potential of these kinds of genetic variants and the nature of the genetic approach.

Neurological and behavioral mutants are useful at two levels. On one level, they can provide tools for disrupting an organism's physiology or anatomy as a simple alternative to direct surgical intervention or drug treatment. In some instances, however, the nongenetic method for achieving perturbations might be equally easy to apply, or even superior for conceptual reasons. For example, there are mutations that result in the absence of certain appendages, such as an insect's wing. One could ask about the effects of these genetic changes on actions that involve this appendage, such as communication by sound production. But what if the mutation had other effects, including less obvious ones on internal anatomy or physiology? Surgical removal of the appendage might be superior to mutational surgery in this kind of experiment.

However, mutations can also be used as disruptive tools in situations where other kinds of experimental intervention are extremely difficult. There are mutations that lead to the removal of a particular cell type from very complicated structures, such as sense organs or parts of the central nervous system, and there are mutant genotypes that result in the removal of particular structures from microorganisms; these cells might have been impossible to treat uniformly by mechanical or chemical means in order to achieve the desired fine surgery. Other kinds of mutations dramatically alter an organism's development at a stage of the life cycle at which direct intervention by the experimenter is very awkward. In this kind of genetic variant, there might be a change in the pattern of cells and tissues whereby cells are misplaced, as opposed to being merely absent. The consequences of such disruption on the development of other cells and the tissues as a whole are then amenable to analysis. There are many examples in which mutations have made it possible to make selective perturbation of an organism's anatomy or function that would have been difficult to obtain by nongenetic means.

Another level at which the usefulness of neurogenetics must be considered relates to the actual function of the genes under study. Through the isolation of mutants, genes can be identified that have profound effects on neural development and function. What products, if any, are specified by these genes? Some of these gene products are already proving to have extremely interesting properties, heretofore unanticipated. It is thus possible that critical comparisons of normal and mutant organisms will lead to the identification and characterization of the molecules involved in assembling nervous systems and controlling their functions.

What, then, are the specific features of nervous systems that have been studied genetically? Many mutations affect the basic physiology of excitable cells at the level of sensory reception, conduction of impulses, and chemical communication between cells. Such mutants have been discovered in microorganisms and invertebrate metazoans on the basis of defective responses to defined stimuli. The sum total of an organism's sensory, neural, and muscular physiology underlies practically all its actions, including behavioral events that involve sophisticated information processing and motor output. "Higher" behaviors have also been studied intensively with genetic variants. These analyses have made use of approaches developed by ethologists and psychologists. Lower invertebrates or microbial organisms studied in detail with respect to physiological abnormalities of genetic origin have also contributed, sometimes suprisingly, to the understanding of higher functions. However, an important proportion of the work on these functions involves studies of evolutionarily advanced metazoans.

Some of the most interesting genetic work on higher behavior concerns modifiable behaviors such as learning and memory. Some behaviors can best be thought of as being primarily controlled by endogenous factors, such as gene action, while others are more likely to be under the influence of outside factors, such as the effects of experience. Controversy has sometimes arisen over which of these types of factors is more critical. We are not concerned with such controversies here; the actions of genes underlie all types of behavior, including mechanisms of learning, since many of the molecules that control the function of neurons are direct products of genes. Thus, we have not concerned ourselves with "nature versus nurture" issues; our interests are far more biological than sociobiological. It is perhaps for these reasons that *Genetic Neurobiology* was chosen as the title of this book instead of *Behavioral Genetics*; a substantial fraction of the latter field is concerned with issues we shall not discuss here because of our biological focus.

After reviewing genetic work on the function of excitable cells and the behaviors they mediate, we shall consider the large area of neurogenetics having to do with the development of the nervous system. Developmental biology and genetics have had extremely important, long-term historical connections at the levels of concepts and experiment. One of the most interesting and perplexing areas in developmental biology is neural development; so it is natural that much attention centers on the genetic control of the assembly of nervous systems.

It seems as if cell differentiation and pattern formation in neural development must be governed in some manner by the action of many genes. Indeed, neurogenetic analysis of development in organisms ranging from primitive invertebrates to higher mammals has identified several of these genes. Whereas we shall not touch on very general issues, such as how gene action is regulated in the development of complex organisms, we shall suggest that some of the exciting progress that continues to be made in developmental genetics has particular connections to the genetic approaches used in studying neural development.

So many problems in neurobiology are being studied genetically that the actual work in this area demands a broad application of principles and techniques. Data are derived from diverse areas, such as anatomy, experimental embryology, biochemistry, molecular biology, neurophysiology, ethology, and psychology. The kind of analysis that must be performed on a given mutant is not necessarily unique to neurogenetics. However, some of the genetic information obtained is unique to neurobiology, particularly with respect to studies on cell differentiation as it relates to chemical neurotransmission or to pattern formation involving the complex assembly of precise neural networks.

Analysis of issues such as these has relied heavily on specialized genetic techniques, such as genetic mosaicism. Studies of cell interactions important in differentiation and pattern formation can often be most effectively analyzed by using neurological mutants augmented by mosaics: individual organisms comprised of more than one genotype in which some cells express a neurological mutation and others are genetically normal. Thus, the existence of an important cell interaction can be established in experiments showing, for instance, that mutant neurons induce changes in target cells of a particular type. Mosaic analysis of neurological variants will figure prominently in our discussions of both invertebrates and vertebrates.

Genetic sophistication has proved to be essential in neurogenetic work, making possible the construction of complex genotypes, such as those required for mosaic production. Ongoing genetic investigations have proceeded in parallel with experiments in the aforementioned areas, such as histology and physiology. There are many examples in which the applications of complex procedures or principles that are purely genetic or molecularly genetic have been critical in furthering the understanding of a neurobiological defect. For instance, it has proved very useful to select more than one mutant allele of a gene of behavioral interest. Examples of the usefulness of multiple alleles come from almost every organism used in neurogenetics. Some studies have relied on the construction of compound genotypes, such as double mutants, to uncover an interaction between two particular genes that were initially understood only by their separate effects. It should be noted, however, that in all of these cases, we are primarily concerned with studies of individually defined genetic variants or, at most, with more complex genotypes that are completely specified. We shall not consider, therefore, at any length the studies on "polygenic control" of behavior (involving strains that differ at a large but unknown number of genes) or on genetic selection of strains that are different behaviorally (and presumably different at many genetic loci). Such complex genetic changes might be tractable by physiological or neuroanatomical analysis, but they are unlikely ever to be comprehensible at the level of gene action.

Other techniques in the realm of "pure" genetics involve procedures such as mapping of behavioral and neurological mutants. It has turned out, again rather surprisingly, that detailed genetic mapping per se has frequently contributed to an understanding of the biological phenomenon controlled by the gene in question. Examples come from cases where neighboring genetic loci have proved to be interesting neurobiologically and from other instances where two separate investigators have identified the same gene, and its precise location, according to totally separate behavioral and biological criteria. Some of the mapping procedures involved not only genetic-recombination experiments but also the use of chromosomal rearrangements or deletions. The ability to produce, recognize, and manipulate these kinds of chromosome aberrations has been valuable in neurogenetic systems ranging from bacteria to mouse. In bacteria, the use of such chromosomal rearrangements has greatly assisted the molecular study of genetic defects in behavior. We shall note several instances of the kind of molecular work that is underway using these and other organisms, including a discussion of the current or potential promise of molecular cloning (recombinant DNA) for identifying the products of genes that control the development and function of the nervous system.

We believe that the broad application of certain genetic tools in neuroscience, as described herein, can be as important as the conventional tools of physiology or biochemistry. This is based on our contention that genetics, neurobiology, and behavior have a closer kindred spirit than one might have imagined. The reason is, in part, historical. Beginning about the time of the "rediscovery" of Mendel's seminal work on the segregation and transmission of genes, many biologists were willing to accept Mendelian genetics as true, but not very interesting or important. After all, single-gene differences at that time had to

do with "trivial" features of organisms, relating to external phenotypes like color or surface texture. As noted by historians of biology
and genetics, it was in fact *essential,* in the early days, that the deduc-
tion of purely genetic principles be deliberately kept separate from the
major efforts aimed at understanding complex organismic development and functioning. In this light, biologists rightly wondered what
were the factors controlling the fundamental features of organisms
that specified their critical attributes. It seemed that these factors
would have to be discovered not by Mendelian approaches but
through the biometrical study of continuously distributed traits, such
as size or "intelligence." Yet it has turned out, in modern genetics and
neurogenetics, that the analyses of single-gene differences, especially
of genes that influence behavior and the nervous system, have identified absolutely essential hereditary factors. We are now ready to
*merge* genetic concepts and technology with studies of complex nervous systems and the complicated, sometimes unique, aspects of the
functions of their cells.

Many of the behavioral and neurobiological mutations are lethal
because they disrupt essential physiological and neurochemical mechanisms. Other mutations perturb the development of organisms,
including the assembly of the central nervous system, in such fundamental ways that the organisms' essential pattern and form are dramatically altered. These mutations, then, do not merely lead to gross
lethality or the simple failure of a tissue to form or function. Instead,
they seem to be involved in generating and maintaining basic species-
specific characteristics. Many other loci identified since the early days
of genetics have also been found to be essential genes implicated in the
special nature of particular organisms.

The complexity of nervous systems, and of the developmental
events that must give rise to them, has led some to believe that neurogenetic analysis is misguided, that the crucial events are too far
removed from the actions of genes. Yet we might ask the simple question, How do we imagine that complex nervous systems, such as those
found in mammals, evolved? Did the process involve some mysterious
changes, over time, in the nature of our cells and tissues, or did it
involve hereditary changes at the level of chromosomes, genes, and
molecules? Clearly, the role that genes play in building nervous systems is not trivial, nor is the manner in which they embody such information obvious.

Our discussion will begin with the basic cellular functions that
underlie a variety of ostensibly simple behaviors. We shall then proceed to more complex phenomena, involving far more than one or a
pair of excitable cells. Throughout, studies on the various organisms
will be woven into the discussion of various neurobiological topics,

which is in contrast to treatments that plow through all the neurogenetic features of each of the different organisms in turn. By organizing the discussion around neurobiological issues, we hope to be able to emphasize the unique knowledge derived from the genetic basis of the experiments.

# 2
# PHYSIOLOGICAL AND
# NEUROCHEMICAL GENETICS

## SENSORY MECHANISMS

### Chemotaxis in Bacteria

Studies of bacteria have made crucial contributions to our understanding of genetics and molecular biology. What could these unicellular prokaryotes possibly have to do with neurobiology? They have much to do with it, in part because bacterial responses to stimuli have been studied for almost 100 years (see Berg, 1975). These ostensibly simple prokayotic cells can exhibit positive and negative chemotaxis, geotaxis, thermotaxis, and phototaxis, as just part of their behavioral repertoire. What are the precise mechanisms of reception and transduction of the received information? And what is the mechanism of adaptation to the stimuli? These fundamental questions are being asked very intensively about a variety of metazoans as well as these prokaryotes. Whereas mechanisms of reception are understood comparatively well in the lower and higher forms, the mechanisms of transduction and adaptation are not well understood. In recent years, chemoreception and chemotaxis have been the bacterial phenomena analyzed most extensively.

Much of the analytical approach has been genetic, and investigators such as Adler, Berg, Koshland, Parkinson, Macnab, and their colleagues are coming close to understanding chemotaxis in *Escherichia coli* and *Salmonella typhymurium* at the molecular level. As might be expected, chemoreception per se in these bacteria yielded first to molecular dissection; transduction and adaptation are now beginning to be understood via the genetic approach. It is hoped that the nature of these bacterial mechanisms will be formally analogous to sensory phenomena in higher organisms.

The concerted efforts toward understanding bacterial chemotaxis began with the isolation of mutants that cannot show normal movements toward potentially attractive chemicals or away from repellents. Most of the useful mutants failed to respond either to specific chemicals or to various combinations of chemicals. The relevant mutants in both categories were otherwise motile, suggesting that these mutations define genes that are specifically involved in the control of reception and transduction, as opposed to being grossly defective in cellular

8

metabolism or structure (Hazelbauer and Adler, 1971; Aksamit and Koshland, 1974; Ordal and Adler, 1974; Hazelbauer, 1975; Kort et al., 1975; Silverman and Simon, 1977b; Goy and Adler, 1977, 1979). The chemotactic mutants began to be characterized behaviorally and biochemically, and the result has been the definition of a pathway of information flow during stimulus transduction (figure 1).

Since the bacterial chemotactic work has been constantly reviewed in recent years (Parkinson, 1977; Macnab, 1978, 1980; Springer, Goy, and Adler, 1979; Koshland, 1979, 1980), an exhaustive account of the genetic, behavioral, and chemical work will not be presented here. Rather, knowledge of sensory mechanisms that has come particularly from experiments on genetic variants will be highlighted.

*Basic Bacterial Behavior*
To monitor bacterial chemotaxis, one can set up a simple concentration gradient of, for instance, attractants such as sugars or amino acids. Normal bacterial will aggregate in a capillary tube containing such molecules (Pfeffer, 1884; Adler, 1969), and the number of individual cells that so respond (figure 2) can be calculated by squirting them out of the tube, growing a colony from each cell, and counting them. A population of mutant cells would not respond in this fashion. However, to find the mutants arising initially in individual cells by chemotactic assay, one can cover a Petri dish with an attractant, observe the cells in the middle of the plate, and then note the normal responders carrying out chemotaxis outward as they use up the attractant. Cells staying in the middle might be nonchemotactic or nonmotile mutants (Adler, 1969). Such cells are isolated, and a series of pure clones of each potentially mutant line is initiated. At this point, behavioral measurements on individual cells from mutant and normal strains can be performed to determine, for instance, whether the mutant has any motility.

These studies of single cells from normal strains have shown that *E. Coli* or *S. typhimurium* cells move in a three-dimensional random

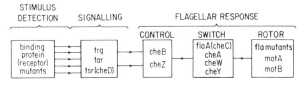

**Figure 1.**
Pathway of information flow during stimulus transduction in bacterial chemotaxis, based on phenotypes of chemotaxis-defective mutants. Mutants are ordered from specific (receptor mutants) to general (motility mutants) with respect to extent of their chemotaxis defects. [Parkinson, 1977]

**Figure 2.**
Photomicrograph showing attraction of *E. coli* bacteria to aspartate. The capillary tube (diameter, $\sim$ 25 m) contained aspartate at a concentration of $2 \times 10^{-3}$ M. Dark-field photography. [Adler, 1969]

"walk" composed of alternating episodes of smooth swimming and tumbling (Berg and Brown, 1972). The motor mechanism involves the clockwise and counterclockwise rotation of flagella covering these cells (see "Behavioral Neurogenetics"). Chemotaxis is mediated by a biasing of the random swimming. Thus, smooth swimming that happens to be up a concentration gradient of attractant or down a repellent gradient will tend to be maintained, in that the probability of tumbling is suppressed.

*Chemoreception*
The way bacterial cells detect differences in concentration of attractants or repellents involves temporal sensing (Macnab and Koshland, 1972). The chemoreceptors at the surface of the cells sense chemical stimuli by monitoring changes in concentration rather than absolute levels of the substances (Macnab and Koshland, 1972; Brown and Berg, 1974). Cells in essentially one place have been shown to alter their behavior at that location, by experimentally induced temporal changes in the concentration of effector chemicals (Macnab and Koshland, 1972). This is a crucial feature of bacterial chemotaxis because it allows the cells to respond to very large changes in levels of stimuli.

When sudden large increases in attractant concentration are made, relatively persistent smooth swimming results. These responses, however, are transient, implying that the sensory system in these cells undergoes adaptation (Macnab and Koshland, 1972; Springer, Goy, and Adler, 1977).

What kinds of mutants have been isolated, and how do they affect the reception of, and response to, stimuli in terms of the details of the phenomena that have been just described? Several mutants (that have normal motility) fail to respond to specific chemicals. Many of the genes defined by these mutations code for the chemoreceptive molecules per se, such as the galactose receptor that resides in the space between the inner and outer membranes of the *E. coli* cell wall (Kalckar, 1971).

The simple identification of a sugar-binding protein, important for chemotaxis, is dramatically aided by the isolation of a mutant lacking both the behavior and the protein (Hazelbauer and Adler, 1971). If binding proteins are involved in the reception of other attractants and repellents, they can be revealed most powerfully by the identification of specific receptor mutants. Mutants can also show rather quickly that reception of more than one related molecule is mediated through the same receptor if a single mutation abolishes the response to all compounds (Hazelbauer and Adler, 1971).

Competition experiments on wild-type cells can also help define how many different receptors are involved in receiving information on a class of compounds. Thus, if large quantities of compound A do not affect the cell's ability to respond to compound B, then one might predict the existence of two different receptors, one for compound A and one for compound B. Blockage of the response to one compound, in the presence of large amounts of another, may also mean that pathways dealing with these substances converge, not at the level of reception, but during subsequent steps of transduction.

In the early days of isolation of receptor mutants, investigators wondered whether metabolism of the stimulant molecules was important, not just for general energy metabolism but also for specific features of the stimulus transduction. This was ruled out by showing, for instance, that a mutant that cannot metabolize galactose can still respond chemotactically to this sugar (Adler, 1969). It was thought that permeases, involved in transport of metabolites into the cell, could be the chemoreceptive binding proteins. Yet permease-minus mutants are still normally chemotactic (Adler, 1969; Ordal and Adler, 1974). However, active transport and chemoreception do have common features, as might be expected. Thus, mutants with altered or absent binding proteins are defective in uptake of the relevant molecules as well as in chemotaxis (Boos, 1969).

Reception of information involving thermotaxis of *E. coli* appears to share components with certain features of the chemotactic mechanisms. For instance, the presence of serine blocks the cell's attraction toward higher temperatures (Maeda and Imae, 1979), whereas other attractants (for example, aspartate) have no effect on thermotaxis. Mutations in a certain gene (*tsr*) that involve the processing of chemotactic information eliminate thermotaxis as well as chemotaxis toward serine. Other evidence from Maeda and Imae (1979) suggests that the receptor is the shared component involved in the responses to these two kinds of stimuli. Still, the *tsr* mutants—isolated according to one criterion and later shown to affect a wholly separate response as well—are one useful way to reveal the "conservative" nature of certain stimulus-response mechanisms, in which one gene product is used in the control of more than one type of behavior.

*Sensory Transduction*

The genetic dissection of the subsequent steps in the pathway (figure 1) seemed, a priori, to possess more mysteries than chemoreception. For the later steps, mutants that are more "generally" chemotactic negative are crucial. Individuals from this kind of strain will fail to respond to a battery of chemicals that are known to be received by different binding proteins. There are three known genes that, when mutant, cause the cells to have defective responses to several chemicals (Mesibov and Adler, 1972; Ordal and Adler, 1974; Springer, Goy, and Adler, 1977). Other generally nonchemotactic strains include mutants in any of about eight or nine other genes that are motile but do not show chemotaxis with respect to all compounds tested (Armstrong, Adler, and Dahl, 1967; Aswad and Koshland, 1974; Collins and Stocker, 1976; Parkinson, 1977; Warrick, Taylor, and Koshland, 1977). All the generally defective mutants can be thought of as affecting parts of the pathway—down the chain from the initial reception events—at (or after) which information from various kinds of stimuli converges. These later steps in the pathway have been defined, at least formally, by the use of the relatively pleiotropic nonchemotactic mutants (see also "Behavioral Neurogenetics").

Genetic mapping and complementation tests of the generally nonchemotactic mutants have revealed the involvement of a relatively small number of genes. If 100 or so genes were involved (instead of approximately 10), then the dozens of mutants isolated would have defined a number of genes approaching this large number. Instead, mutations that are newly discovered proved to be extra alleles in genes already known to control chemotaxis. Moreover, when a mutation is discovered that might suppress the effects of a chemotactic mutation in another gene, the former mutation frequently proves to be in a gene that itself was previously known to control chemotaxis (for example, Parkinson and Parker, 1979). This finding further supports the idea that the bacterial maps are becoming "saturated" with respect to knowledge of the genes controlling these stimulus-transaction pathways. In addition, the gene interactions implied by the induction of suppressor mutants have their own intrinsic interest.

The steps of the chemotactic pathway in bacteria will be discussed in the order in which they may occur, inferred from the genetic, behavioral, and, in some cases, biochemical evidence on mutations disrupting these steps. It must be kept in mind that this discussion essentially presents a model (figure 3; Parkinson, 1977) and that the number and order of steps have not been completely confirmed.

The step in signaling after reception may involve the products of the genes *tar* (taxis to aspartate and certain repellents and attractants), *tsr* (taxis to serine and certain repellents and attractants), and *trg* (taxis to ribose and galactose). Mutations in these genes were isolated on the basis of pleiotropic abnormalities, implying that these three genes are at a relatively convergent step in the pathway. The *tar* mutants are not attracted to particular sugars and amino acids, and they fail to respond to a particular set of repellents. The *tsr* mutants have essentially a complementary phenotype, that is, respond normally to compounds that do not influence *tar* mutants; but the *tsr* mutants are not attracted to or repelled by the particular compounds to which *tar* mutants do respond. The *trg* mutants abolish responses to ribose and galactose, but chemotaxis to these sugars is not affected by either *tar* or *tsr* mutants. The *trg* gene codes for what was earlier called the "shared signaling element" that controls the second step in the ribose-galactose pathway of chemotaxis. This shared element was postulated on the basis of experiments in *Salmonella* (Strange and Koshland, 1976) showing that saturation of the ribose-binding protein blocks chemotactic responses to galactose, though the receptors for these sugars are separate entities. Since these two receptors are among the best characterized in bacteria, the system involving these receptors and the next step of the pathway should allow molecular experiments on the interactions between the receptor molecules and the molecules controlling the transduction events immediately subsequent to reception (Wang and Koshland, 1980).

There is, in fact, extensive molecular-genetic information on the functions mediated by *trg* and, especially, *tar* and *tsr* genes (Kondoh, Ball, and Adler, 1979). The intermediate step in chemotactic control, mediated by the products of these genes, is thus yielding to analysis at a concrete level, in addition to the level of formal pathway dissection. Moreover, the studies on molecules affected by these three genes indi-

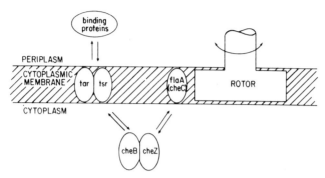

**Figure 3.**
A model of stimulus transduction in bacterial chemotaxis. The *flaA* (*cheC*) component may be a membrane protein that interacts with the flagellum to control its direction of rotation. [Parkinson, 1977]

cate that they are involved in both chemotactic transduction and adaptation to chemotactic stimuli (Goy, Springer, and Adler, 1978).

The discovery of biochemical correlates of information on *tar, tsr,* and *trg* began with the observation that methionine is continuosly required during chemotaxis in *E. coli* (Adler and Dahl, 1967) and *Salmonella* (Aswad and Koshland, 1974). Metabolic mutants that cannot synthesize this amino acid when deprived of it lose their ability to tumble and thus change their direction of swimming (Springer et al., 1975). Such mutants also cannot adapt to stimuli that suppress or enhance tumbling (Springer, Goy, and Adler, 1977).

Methionine is involved in donating a methyl group to other molecules through the intermediary of *S*-adenosyl methionine. Methyl-accepting proteins identified in the cytoplasmic membrane of bacteria are involved in chemotaxis and are called methyl-accepting chemotaxis proteins (MCP) (Kort et al., 1975). Some experiments of Silverman and Simon (1977 a,b) involving the molecular cloning of the *tar* and *tsr* genes, each in separate phage-lambda-transducing viruses, have shown that these two genes each code for membrane proteins of about the same molecular weight as the MCP polypeptides. DeFranco and Koshland (1980) used a cloned *tar* gene from *Salmonella* (in a lambda phage vector) to show that the *tar* gene product from this species can be methylated and can lead to normal chemotaxis in *tar* mutants of *E. coli*. Using the cloned gene, these authors also showed that the *tar* gene product is multiply methylated (accepting at least four methyl groups) after exposure of the bacteria to attractants. Similar results were found by using a two-dimensional analysis of methyl-accepting tryptic peptides (Chelsky and Dahlquist, 1980).

Either *tar* or *tsr* mutants produce reduced amounts of the MCPs; double mutants make very little MCP, exhibiting only the residual molecules (of about the same molecular weight as *tar* and *tsr* gene products) whose methylation can be stimulated by ribose and galactose. Finally, the triple mutant has no detectable MCP (Kondoh, Ball, and Adler, 1979).

To connect these biochemical facts with behavioral ones is not difficult. Increases in attractants that induce smooth swimming also stimulate increases in methylation of MCP. Increases in repellent levels cause concomitant increases in tumbling frequency associated with reduced methylation of MCPs (Silverman and Simon, 1977a; Springer, Goy, and Adler, 1977)

Levels of methylated MCPs are also involved critically in sensory adaptation, since adaptation to increases in attractant is correlated with methylation of MCPs and adaptation to decreases in attractant is correlated with demethylation of these proteins (Springer, Goy, and Adler, 1977). Levels of methylation of the MCPs are controlled by a balance between methylase and demethylase activities. Using a cell-free extract of *E. coli* and measuring either methylation or released methanol, Kleene and coworkers (1979) found that attractants enhance methylation and inhibit demethylation, while repellents do the reverse. A mutant of the *cheB* gene proved to be useful in this study because it had no demethylase activity (Toews et al., 1979; Kleene, Hobson, and Adler, 1979). The *cheB* mutant was initially discovered behaviorally as an incessant tumbler. Another mutant, *cheX*, also discovered on the basis of abnormal swimming behavior (smooth swimming, unless exposed to repellents, which still leads to abnormally long response times; see Parkinson, 1977) turned out to be defective in methylation reactions (Toews et al., 1979; Hobson and Adler, 1980).

The connection between levels of methylated MCPs and both transduction and adaptation has not been explicitly worked out yet; but more information is becoming available on these matters through studies of mutations that affect later steps in the transduction pathway (see "Behavioral Neurogenetics").

Double mutants, lacking both MCP-1 (coded for by *tsr*) and MCP-2 (from *tar*), also have low frequencies of tumbling, implying that the methylated forms of MCP molecules are important in the operation of the tumble-generating machinery (Silverman and Simon, 1977a). The double mutants still show some chemotaxis to ribose and galactose since the *trg* gene product (MCP-3) is still active.

Additional effects of the *tar* and *tsr* genes have been discovered through the isolation of a mutant *E. coli* that is attracted to certain compounds that are normally repellents (for example, acetate) and repelled by the normal attractant $\alpha$-aminoisobutyrate (Muskavitch et al., 1978). This "reversed" mutant was soon found to be an allele of *tsr*. Since the MCP-1 system is defective in *tsr* mutants, it was thought that the reversal mechanism utilizes the remaining MCP-2 system (specified by the *tar* gene). Indeed, in the *tsr* mutants, MCP-2 shows increases in methylation when repellents are added to the cells, and methylation decreases when attractants are added (Springer, Goy, and Adler, 1977; Muskavitch et al., 1978).

The use of these mutants has revealed that changes in concentration relating to the supposedly separate classes of compounds (that is, separate for the *tar* and *tsr* signaling pathways) affect both MCP-1 and MCP-2 molecules in the cell's membrane simultaneously. Adler and colleagues (Muskavitch et al., 1978) speculated that this phenomenon may be of adaptive significance. The cell may at certain times of nutritional stress "want" to use a normal repellent (for example, acetate) as a carbon and energy source. In any event, these data further stress the interaction between two kinds of methyl-accepting proteins involved in chemotaxis.

Involvement of the genes controlling MCPs in chemotaxis is additionally suggested by the esoteric but interesting finding that certain generally nonchemotactic mutants—in a gene originally called *che-1*—are dominant to the normal gene. These mutants are unable to tumble. When one induces additional mutations in this gene in order to abolish the dominance, then a standard *tsr* phenotype also results (Parkinson, 1977). These findings suggest that the normal allele of the "*che-1* gene" (that is, *tsr*⁺) produces an inhibitory protein and that the original dominant mutation resulted in this protein being locked in a tumble-inhibiting configuration. The second mutation might abolish the action of this gene altogether, allowing normal chemotaxis only for the compounds to which the standard *tsr* mutants will respond. This series of findings reinforces the fact that *tsr* is involved in passing along chemotactic signals to the mechanisms that directly control smooth versus tumble swimming (see figure 3).

Cyclic GMP (cGMP) seems to be involved in the intracellular signaling for the chemotactic response. Black and coworkers (1980) showed that the addition of attractants to *E. coli* leads to transient increase in cGMP levels. Addition of repellents leads to a transient decrease in levels of this nucleotide. If neither attractants nor repellents are added, addition of cGMP causes smooth swimming; reducing cGMP induces tumbling. Mutants that tumble incessantly, such as *cheB,* can be induced to swim smoothly by addition of cGMP. Furthermore, mutants with abnormal swimming behaviors, such as *cheB, cheX,* and *flaI,* have levels of cGMP production that are appropriately altered in the context of their behavioral abnormalities. Black and collaborators (1980) also showed that cGMP inhibits demethylation of MCPs and thus affects adaptation; moreover, adaptation feeds back on cGMP levels because mutants with high levels of methylation (that is, resulting from expression of the *cheB* demethylation mutant) have decreased production of cGMP. The reverse is true with *cheX,* which is defective in methylation. The *flaI* mutant, which expresses none of the chemotaxis-specific proteins, also has increased production of cGMP. Thus, it is likely that methylated MCPs actively inhibit cGMP formation.

In summary, it appears that the excitation and adaptation of bacterial chemotaxis can be roughly modeled as follows: Receptors bind attractants, leading somehow to a rapid increase in cGMP. High cGMP levels cause a prolonged period of smooth swimming. Simultaneously, increased cGMP inhibits demethylation of the MCPs, thus increasing the level of MCP methylation, which in turn inhibits cGMP production. Levels of cGMP decline as a result, and the organism reverts to an unexcited-walk state. A similar, but essentially opposite, scenario can be postulated with respect to repellent-stimulated behavior.

Nearly all of our understanding of how bacteria receive chemical stimuli and begin to process this information has come from the use of genetic variants. The uncovering of the mechanisms controlling these early stages of the behavioral pathway are interesting to more than microbiologists. It is possible that cellular phenomena underlying sensory transduction and adaptation in metazoans have analogies to the molecular events now being uncovered in the bacteria. Receptor-ligand interactions are one possible analogy; another one is the link between events at the level of reception and other molecular processes such as protein modification. For example, there is much current work on phosphorylation of membranous proteins in vertebrate cells. These proteins appear to play an important role in the response of certain neurons to signals from other cells and may also be related to changes in levels of cyclic nucleotides (for example, Greengard, 1976). Phosphorylation and its relevance to cyclic nucleotide regulation can, without difficulty, be viewed as analogous to the methylation events and the recently discovered role that cyclic GMP appears to play in intracellular signaling in bacteria. Moreover, chemotaxis per se is important for certain cells in higher forms, such as slime molds or white blood cells in mammals. The possible involvement of methylation of some component in these cells is suggested by the work of several groups (O'Dea et al., 1978; Pike, Kredich and Snyderman, 1978; Mato and Marin-Cao, 1979). Bacterial behavioral genetics, then, seems very much in the mainstream of biology in general and neurogenetics in particular.

### Chemosensory Behavior in Eukaryotic Organisms

*Paramecium*
A sensory pathway analogous to that for chemotaxis in bacteria is also found in the unicellular protozoan *Paramecium tetraurelia*. This organism, which has many of the properties of membrane excitability normally associated with neurons, exhibits attraction to, and repulsion from, a variety of chemical stimuli. This behavior was exploited initially for the purpose of isolating mutants with defects in membrane excitability (reviewed by Kung et al., 1975; Kung, 1979), but, subsequently, the behavior itself was studied in some detail by making use of several of the mutants affecting excitability. In contrast to the bacterial studies, relatively few portions of the pathway have been defined in *Paramecium,* but the genetic studies have resolved the chemosensory response into two distinct behavioral components and have helped to establish the role of membrane potential in regulating the choice between them.

Paramecia move about by the coordinated beating of the many cilia that cover their surface. They alter their swimming direction by means of an avoiding reaction, in which they abruptly reverse the orientation of their cilia and then set off in the new randomly oriented direction that they have assumed. This avoiding reaction may be considered to be roughly analogous to the tumbling of bacteria; however, the behavior of paramecia toward attractants and repellents cannot be sufficiently explained simply by the frequency of avoiding reactions. The first clue to this came from Van Houten's (1977, 1978) studies of a mutant (*d4-530*) that failed to be attracted by sodium acetate. This chemical is normally attractive to wild types in a T-maze assay that tests the cells' choice between a control solution of NaCl and the test solution (Van Houten, Hansma and Kung, 1975). Nor did the mutant behave indifferently to acetate; instead, it was repelled by it. Upon closer examination, the mutant was seen to exhibit fewer avoiding reactions than the wild type, while at the same time its swimming velocity was significantly increased. Thus, it became apparent that the velocity of swimming was as important as avoiding reactions to the cell's ultimate response.

Van Houten (1978) pursued this idea further by examining the mode of wild-type response to a number of different attractants and repellents. It was found that the cells had a repertoire of four responses: two types of attraction and two types of repulsion, all modulated by the frequency of avoiding reactions and the magnitude of the swimming velocity. Both classes of attractants (type I attractants, such as acetate) and one class of repellents (type II repellents, such as hydroxide ions) caused a decrease in avoiding reactions *and* an increase in swimming velocity. What makes the one attractive and the other repulsive is that acetate causes a significant decrease in avoidance but only a mild increase in velocity, while hydroxide ions cause a comparable decrease in avoidance but an overriding increase in swimming velocity. Conversely, attractants, such as barium ions (type II attractants), and repellents, such as quinidine (type I repellents), work by increasing the frequency of avoiding reactions and decreasing velocity, again in varying degrees that determine the choice of response.

Several of the behavioral mutants proved to be instrumental in sorting out the contributions of each mode of response by their selective elimination of one of the two components. Thus, mutants that never show avoiding reactions, called pawns, failed to show either of the types of attraction (type II) or repulsion (type I) that require avoiding reactions, but they were altogether normal in the response that routinely causes avoiding reactions to disappear, type II repellents. Another mutant, Fast-2, was of particular interest because it responds abnormally to NaCl, the salt used as the "control" solution in the T-maze test (Van Houten, 1978). Whereas the wild type is induced to make avoiding reactions when exposed to NaCl, Fast-2 does not. Therefore, when Fast-2 exhibited attraction or repulsion to a stimulus relative to NaCl, the contribution of swimming velocity by itself could be determined. It was found, predictably, that in the absence of avoiding reactions, the degree of repulsion depended upon the magnitude of the swimming velocity. The results with Fast-2 are particularly convincing since the phenotype of this mutant is conditional in that it shows normal attraction and repulsion as long as NaCl is not used as the control solution. Thus, the contribution of swimming velocity to the responses in NaCl is not likely to be due to a general disorganization of the chemosensory machinery caused by the mutation.

On the face of it, the antagonism between avoiding reactions and swimming velocity seems difficult to comprehend, but Van Houten made sense of it by correlating the behaviors with previously known effects of membrane potential on swimming behavior (reviewed by Kung et al., 1975). Specifically, avoiding reactions are increased as membrane potential increases (depolarizes), and swimming velocity increases as membrane potential decreases (hyperpolarizes). On this basis, Van Houten predicted (1978) and then demonstrated (1979) that the four classes of response correspond to four different ranges of membrane potential (figure 4), where each interval of potential correlates with the particular mixture of avoiding reactions and swimming velocity seen for each stimulus: (1) large hyperpolarizations give type II repulsion; (2) moderate hyperpolarizations give type I attraction; (3) moderate depolarizations give type I repulsion; and (4) large depolarizations give type II attraction. Furthermore, these findings

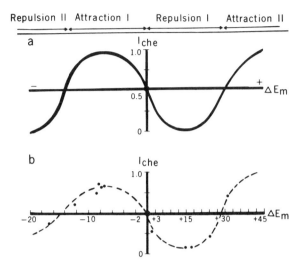

**Figure 4.**
(*a*) Graphical description of membrane potential control of chemokinesis in *Paramecium*. Changes of membrane potential ($\Delta E_m$) from control (at origin) is plotted against the index of chemokinesis: $I_{che} > 0.5$ indicates attraction; $< 0.5$ indicates repulsion. As chemical stimuli change $E_m$ relative to control, animals will be attracted or repelled, depending on the magnitude and direction of the $E_m$ change. (*b*) Data on membrane potentials of cells in test and control solutions, plotted as $\Delta E_m$ produced by the attractant or repellents versus $I_{che}$. Scale of $\Delta E_m$ is different for depolarizing and hyperpolarizing stimuli. [Van Houten, 1979]

correlated well with the behavioral phenotype of the mutants, in terms of their defects in membrane permeability and excitability. Specifically, the absence of action potentials in paramecia was consistent with their failure to show the kinds of attraction (II) and repulsion (I) that require membrane depolarization. Similarly, the response of *Fast-2* to NaCl, a hyperpolarization, was consistent with its faster swimming and lack of avoiding reactions under these conditions (see "Nerve Impulses and Ionic Channels").

These studies have shown that the behavioral responses of *Paramecium* to various chemicals can be understood in terms of graded effects of the stimulus on the cell's membrane potential. This step corresponds approximately to the hypothesized tumble regulator of bacteria, but the analogy holds only insofar as its position in the pathway is concerned, that is, as the common process into which various sensory receptor pathways feed. Membrane potential per se is almost certainly not a significant regulatory factor in bacterial chemotaxis (Miller and Koshland, 1977; Snyder, Stock, and Koshland, 1981), although suggestions as to its involvement have been made [Szmelcman and Adler, 1976).

Earlier steps in the *Paramecium* pathway have not been analyzed; but it is possible to imagine how different classes of receptors could give rise to characteristic changes of membrane potential, by analogy with the differential physiological responses to various chemicals recorded from chemosensory cells of insects (Dethier, 1969). The late stages in the pathway concern the cell's general motor output. Studies of the ciliary apparatus have established the link between depolarization and avoiding reactions by showing that calcium entry during action potentials causes ciliary reversal (see "Nerve Impulses and Ionic Channels"), but nothing is known about the control of swimming velocity. The study of chemosensory pathways in *Paramecium* is not as advanced as in bacteria, but the tractability of the lower ciliate for genetic, biochemical, and physiological studies makes it suitable for analysis of sensory processes that are likely to have direct application to those of higher organisms.

*Nematodes*
Chemotaxis in nematodes may have a fundamental similarity to that in bacteria and in paramecia. Dusenbery (1980) found that a purely temporal change in chemical stimulation can cause a large change in the probability of backward swimming of these roundworms and that these altered probabilities "adapt back" to the basal level in about 1 min, depending somewhat on the intensity of stimulation. Bouts of reversal are associated with changes in the direction of forward motion. Thus, it appears that in nematodes, too, chemotactic behavior is mediated by a random walk strategy, in which probabilities of tumbling are affected both by temporal changes of chemicals in the environment and by adaptation (Dusenbery, 1980). With weaker stimulation, however, another form of chemotaxis may occur in which the animal swims directly up the concentration gradient of the attractant, with only slight and nonrandom changes in direction, the probability of which is not affected by changes in stimulus intensity (Ward, 1973).

Several mutants of chemotaxis have been isolated in *Caenorhabditis elegans,* the free-living soil nematode used in virtually all neurogenetic experiments in roundworms. The mutations found to date affect response to various cations, anions, cyclic AMP (cAMP), pyridine, and amino acids such as tryptophan (reviewed by Ward, 1977, 1978). Two of these mutants, *che-2* and *che-3,* were found to have anatomical defects in the ciliated endings of all their anterior sensory neurons (Dusenbery, 1975; Lewis and Hodgkin, 1977). In other mutants, however, no clear correlation could be found between the class of defective cell and the stimulant that was ineffective in eliciting a response.

Despite the difficulties in interpreting the effects of some of the nonchemotactic mutations, studies of other such variants in nematodes have revealed further similarities between these animals and both bacteria and paramecia. About half of the nematode mutants isolated as thermotaxis defective (Hedgecock and Russell, 1975) were found also to be chemotaxis defective (Dusenbery and Barr, 1980). Moreover, about half of the mutants selected as showing no chemotaxis turned out not to be able to exhibit thermotaxis either (Dusenbery, Sheridan, and Russell, 1975). In *Paramecium,* pawn mutants are also both chemotactic and thermotactic (Hennessey and Nelson, 1979), and, in bacteria, recall that the mechanisms controlling chemotaxis and thermotaxis appear to be strongly coupled (Maeda and Imae, 1979).

*Drosophila*

Chemosensory pathways have been studied in *Drosophila melanogaster* by isolating mutants defective in their response to specific gustatory stimuli (Isono and Kikuchi, 1974; Falk and Atidia, 1975; Rodrigues and Siddiqi, 1978; Tompkins et al., 1979; Siddiqi and Rodrigues, 1980). These mutants have altered responses to NaCl, quinine sulfate, and various sugars, but, for most of them, we have only suggestions regarding where they exert their effect in the chemosensory pathway.

At the most peripheral point in the pathway, the reception of the stimulus, two of the mutants affecting sugar response may have identified an actual receptor or some component of sugar-specific sensory hairs. The $gustA^{xl}$ mutant of Rodrigues and Siddiqi (1978) responds very poorly to sucrose as compared to the wild type, while its responses to NaCl and quinine are normal. This defect was inititally observed in a simple behavioral test in which the fly was stimulated to extend its proboscis after a drop of the stimulant is applied to its tarsus (foreleg) or labellum (mouth). In extracellular recordings from sensory hairs on the fly's labellum, the mutant $gustA^{xl}$ was found to lack the 0.8-mV spikes characteristic of the sucrose response (Rodrigues and Siddiqi, 1978; Siddiqi and Rodrigues, 1980).

More specific information concerning sugar receptors in *Drosophila* and the putative genetic identification of one of them have come from physiological studies of a mutant's response to a series of sugars with related molecular structures (Isono and Kikuchi, 1974; Tanimura, Isono, and Kikuchi, 1978). Labellar hairs of the mutant *126BO4* responded with normal impulse frequency to fructose, but showed a large reduction in the response to glucose. Tests of compounds with related structures indicated that the mutant gave a range of diminished responses to those compounds containing exposed $\alpha$-D-glucopyranosyl groups. The authors noted that the specificity of the abnormal response resembles the substrate specificity of the enzyme $\alpha$-glucosidase, which has been previously hypothesized to be a sugar receptor in insects (Hansen, 1969). This gene is thus a good candidate for specifying a receptor protein. Although no test of allelism has been done, the two mutants gustA^xl and 126BO4 do not appear to be in the same gene. The first has been mapped to the X-chromosome (Siddiqi and Rodrigues, 1980), while the other, although unmapped as yet, does not behave as though sex linked (Isono and Kikuchi, 1974).

Lesions of the chemosensory pathway subsequent to chemoreception have been suggested by several mutants isolated by Tompkins and coworkers (1979), using a countercurrent device that quantitatively measures chemotactic behavior. Mutants in one X-linked gene, gusB showed general ataxis in response to either NaCl or quinine and were also nonresponsive to sucrose in tests of proboscis extension. Another mutant, gusE^N13 was generally mistactic to both NaCl and quinine and also showed reduced proboscis extension to surcrose. By analogy with bacterial chemotaxis, these mutants may affect stages in the process after the inputs from specific receptor cells have converged. Alternatively, they may affect processes for which there is no bacterial analogy, such as developmental events common to more than one type of receptor cell. This is suggested by the fact that some of these gustatory mutants are temperature sensitive for their behavioral abnormalities (Rodrigues and Siddiqi, 1978; Tompkins et al., 1979). Moreover, certain mutants have been found to have temperature-sensitive periods only during development stages. One such mutant is gusA^Q1. This conditional mutant is sensitive to cold temperatures, and the temperature-sensitive period for aberrant response to quinine is during late embryogenesis (Tompkins, 1979). Thus, some effect of this gene at a time when the embryonic brain, for example, is differentiating prevents the animals from responding normally to quinine all through the life cycle, even though the chemoreceptor cells of the adult do not begin to form until days after the temperature-sensitive period. The mutants with developmental defects may have central lesions, as opposed to purely physiological defects in peripheral cells.

Other mutants related to chemosensory behavior and with effects on more than one developing stage of Drosophila are those isolated on the basis of olfactory criteria. One such mutant, the smell-blind (sbl) variant, was found in a mutagenized strain tested for learning ability (Aceves-Pina and Quinn, 1979). Mutant adults were apparently unable to sense or respond to a variety of different volatile compounds. The sbl mutation also affects larval olfaction; thus, the suggestion can be made that the defect is in the central processing of odor cues, since it might be that olfactory receptors are quite different in the larvae and adults.

Other olfactory mutants in *Drosophila* have been isolated in a more systematic fashion (Rodrigues and Siddiqi, 1978; Rodrigues, 1980). Some have been found to respond aberrantly only to certain odors, whereas other mutants are defective with regard to several different compounds. An individual from the former class of mutants may have a particular peripheral defect in an olfactory receptor, whereas the latter type may be more centrally disrupted. Whatever the sites of defects in these mutants, they and *sbl* have been useful in assessing the importance of olfaction in higher behaviors, such as courtship. It should also be pointed out that the olfactory-deficient mutants are normal in their taste responses, and the taste mutants discussed above respond to odors normally. Finally, from a genetic standpoint it is important to note that more than one mutant allele of most of the separate gustatory and olfactory genes has been independently isolated (Rodrigues and Siddiqi 1978; Tompkins et al., 1979). Some of the taste-defective mutants found by different methods in these two studies have subsequently been shown to be allelic (V. Rodrigues, unpublished). All of these mutations have been isolated on the X-chromosome (since that is where they have been looked for); so this portion of the fruit fly's genome may be nearly "saturated" with respect to the identification of genes important for these modes of chemosensory behavior (see discussion of bacterial chemotactic mutants and of visual mutants in *Drosophila*).

## Phototransduction in *Drosophila*

The transduction of light stimuli into neural signals by cells of the visual system is a problem that has received considerable attention in neurogenetic studies of sensory physiology in higher organisms. The study of mutants of the visual system in *Drosophila* has been facilitated by the ease of isolating mutants on the basis of either a failure to exhibit phototaxis in behavioral test or abnormalities in the electroretinogram (ERG) (Pak, 1975, 1979). Moreover, the relatively large cell bodies of the photoreceptors in *Drosophila* have made possible intracellular recording, thus permitting the considerable amount of physiological information collected from photoreceptors of this and other organisms to be applied to the genetic variants.

*Genetic Demonstration of Different Visual Pigments*
The first step in phototransduction, light absorption, is mediated in *Drosophila* (as in other organisms) by rhodopsin. The principal rhodopsin in *Drosophila* resides in six outer photoreceptors of each ommatidium (eye facet). This visual pigment absorbs maximally at 480–490 nm and photointerconverts with a thermally stable intermedi-

ate, metarhodopsin, with maximal absorption at 550–580 nm (Ostroy, Wilson, and Pak, 1974; Pak and Lidington, 1974). An analysis of which pigments are in the different kinds of photoreceptors involved a study of mutations that specifically eliminate particular subclasses of these cells (that is, all the outer cells from each facet or one kind of inner photoreceptor from each facet). Thus, Harris and coworkers (1976) showed that the *outer cells* contain a blue-absorbing pigment (note the 480–490 nm maximum) with an ultraviolet-absorbing secondary peak. The more distal *central cell* contains a pigment absorbing maximally in the ultraviolet (UV). Without the use of the mutation (called sevenless) that eliminates a central cell from each facet, the existence of this specialized UV receptor might have remained hidden.

*Phototransduction Mutants*
The steps in visual excitation that immediately follow light absorption are the major objects of most of the genetic studies on phototransduction. The ultimate result of the process, the receptor potential, is essentially the same as that found in many invertebrates: a depolarization of the membrane, graded with light intensity and mediated primarily by sodium ions (Millecchia and Mauro, 1969; Wilcox, 1980). The photoreceptor potential is composed of a summation of many discrete potentials known as quantum "bumps," each of which represents the response to one quantum of light (for example, Fuortes and Yeandle, 1964; Dodge, Knight, and Toyoda, 1968; Wu and Pak, 1975).

No mutants in *Drosophila* have yet been analyzed that definitely identify the genes for the protein component of rhodopsin, chromophore synthesis, or the membrane channels mediating quantum bumps. However, two mutants have been studied that formally identify components of the hypothetical intervening process. These are noreceptor-potential-A (*norpA;* Hotta and Benzer, 1970; Pak, Grossfield, and Arnold, 1970; Heisenberg, 1971) and transient-receptor-potential (*trp;* Cosens and Manning, 1969).

The *norpA* mutants were isolated in deliberate screens for flies with visual defects, either in phototactic or optomotor responses. Most such defects would be thought to result in abnormal function (or structure) of the major photoreceptors in the compound eyes of the adult. However, *norpA* mutations have subsequently been found *also* to disrupt the function of larval photoreception (with respect to visually triggered choice of pupation sites; Markow, 1981) and to cause abnormalities of photoreception by the ocelli in adults (with respect to physiological recordings from these simple eyes; Reichert, and Stark, 1978).

Flies expressing a *norpA* mutation are not defective in the primary

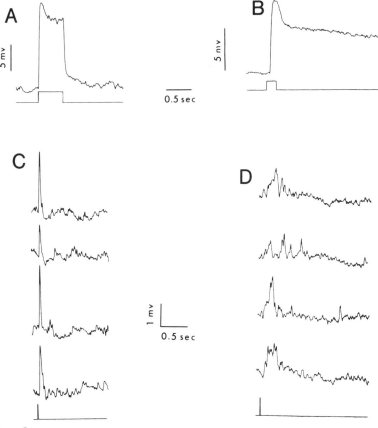

**Figure 5.**
Intracellular recordings of photoreceptor responses obtained from wild-type (*A, C*) and *norpA*[H52] (*B, D*) *Drosophila* at room temperature. (*A*) Response of a wild-type photoreceptor to a 0.5-sec, 475-nm stimulus of −1.5 log relative intensity. (*B*) Response of a *norpA*[H52] photoreceptor to a 0.2-sec, 500-nm stimulus (log $I$ = 0). (*C*) A series of responses of the wild-type photoreceptor to 10-msec, 475-nm flashes (log $I$ = −4.0). (*D*) A series of *norpA*[H52] responses to 10-msec, 500-nm flashes (log $I$ = −1.5). Somewhat higher stimulus intensities were used for the mutant to compensate for its higher threshold. The highest available intensity (log $I$ = 0) corresponds to illuminance of $10^{11}$–$10^{12}$ photons cm$^2$ sec$^{-1}$. [Pak et al., 1976]

stage of response to light because they develop normal levels of rhodopsin, which undergoes normal photointerconversions (Pak et al., 1976; Ostroy, 1978; Pak, 1979). Examination of individual quantum bumps was made possible in the case of the blind *norpA* mutants by using an allele that did not completely eliminate the photoresponse, as many *norpA* alleles do (Pak, 1975), that is, the temperature-sensitive mutant *norpA*$^{HS2}$ (Deland and Pak, 1973). Alleles of the *norpA* gene with residual function, such as *norpA*$^{HS2}$ at its permissive temperature, produce receptor potentials that persist for an abnormally long time after the end of the light stimulus (figure 5; Pak et al., 1976). Intracellular recordings from this mutant clearly showed that the amplitude and time course of individual bumps were normal, but that the timing of their appearance following a flash of light was much less synchronous than in the wild type (Pak et al., 1976). Thus, whereas extreme *norpA* alleles fail to produce bumps altogether, residually active alleles are apparently impaired in the precision with which they respond to light by producing bumps.

Flies expressing a *trp* mutation, which like *norpA* does not affect rhodopsin levels, behave as if they are blind under high light intensity. In order to examine bump parameters influenced by *trp,* Minke and coworkers (1975a) quantitatively analyzed the fluctuations of voltage noise recorded in photoreceptors, a method for estimating the amplitude, time course, and rate of production of quantum bumps for conditions under which individual bumps cannot be resolved. This indirect approach was necessitated by the fact that at low light intensities, where bumps are still resolvable, the *trp* phenotype is normal. The mutant phenotype does not appear unless relatively bright, prolonged illumination is given, and even then the response is normal for the first second or so, after which it decays to baseline or near baseline despite continued illumination (figure 6) (Cosens, 1971; Minke, Wu, and Pak, 1975a). It was important, therefore, to be able to estimate bump parameters at different light intensities and at times after the decay of the initial response to bright light. The result was that the *trp* mutant produced normal quantum bumps in dim light, but, as the intensity was raised, the rate at which bumps were produced was not maintained at the same level as in the wild type (Minke, Wu, and Pak, 1975a). Thus, it appears that the *trp* mutant responds initially in a normal way to a bright light, but that the mutant cannot continue to generate the appropriate number of quantum bumps beyond the first second or two of exposure to light.

The parameters affected by these two mutants can be considered in the context of current models for the intermediates of phototransduction (reviewed by Cone, 1973). According to the models, the opening of membrane channels in a quantum bump is the result of their interaction with a "transmitter" substance within the confines of the cell, whose liberation or production is triggered by rhodopsin. Of possible relevance to this model are some recent biochemical studies that have identified a variety of light-induced enzymes in vertebrate photoreceptors (for example, Wheeler and Bitensky, 1977; Liebman and Pugh, 1979). Therefore, the $norpA^+$ gene product may be involved in the coupling between rhodopsin and the internal "transmitter release" mechanism, which is perhaps completely uncoupled in some alleles and marginally coupled in others, such as $norpA^{H52}$. The $trp^+$ gene product, on the other hand, may be required for recycling "spent" transmitter or for regeneration of the transmitter "producer," so that, in the $trp$ mutant, the availability of the substance is quickly reduced. These mutants may ultimately help to validate or disprove the internal transmitter model of phototransduction.

Another gene implicated in the intermediate steps of phototransduction causes a light-induced degeneration of photoreceptors when mutant. This is retinal-degeneration-B ($rdgB$; Hotta and Benzer,

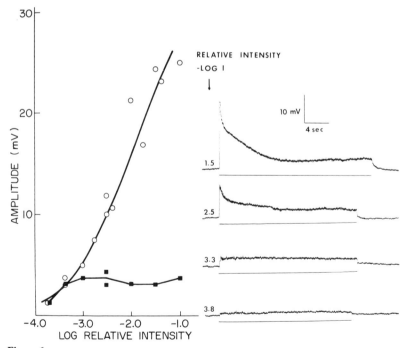

**Figure 6.**
Intracellularly recorded responses from photoreceptors of the white-eyed, $trp$ double mutant of *Drosophila* to various intensities of monochromatic 540-nm green light. The unattenuated illuminance was $3.6 \times 10^{-14}$ photons cm$^{-2}$ sec$^{-1}$ at 540 nm. The lines below the responses indicate the duration of the light stimulus. [Minke, Wu, and Pak, 1975a]

1970). The strongest evidence that this mutant is involved in primary transduction process, as opposed to some general metabolic function required for the maintenance of the cells, is the interaction of alleles of *rdgB* with mutations of the *norpA* locus (Harris and Stark, 1977). Specifically, severe alleles of *norpA* were found to suppress degeneration in double mutants with *rdgB* alleles, implying that the receptor potential per se might bring on degeneration in the presence of the *rdgB* mutation. However, the interaction was shown to be even more intimate with the demonstration that one allele of *norpA,* induced and isolated specifically as a suppressor of *rdgB^{KS222}*, called *norpA^{suII},* had a *normal* receptor potential but was able, nonetheless, to prevent receptor degeneration in the double mutant. Moreover, this particular interaction was allele specific; that is, *norpA^{suII}* failed to suppress degeneration in a different allele of *rdgB* (*KO45*), suggesting a direct interaction of the two gene products in the double mutant *norpA^{suII} rdgB^{KS222}* and, implicitly, in the wild type.

The defects in *rdgB* do not reside in the ability of rhodopsin to photointerconvert with metarhodopsin, as measured in dark-grown flies (Harris and Stark, 1977). However, it appears that some step in the photoresponse is unable to "recover" from the events triggered by a light stimulus of appropriate wavelength and sufficient strength to convert a net amount of rhodopsin to metarhodopsin. A single short exposure (1–2 min) of dark-grown mutants to such a stimulus renders the photoreceptors permanently inactive. The response of wild-type photoreceptors to the same stimulus is to become depolarized and unresponsive to subsequent stimuli of the same (blue) wavelength. But this condition (known as the prolonged depolarizing afterpotential, or PDA) is not permanent and can be instantly reverted to the initial, responsive state by a flash of orange light (Minke, Wu, and Pak, 1975b). The *rdgB* mutants, on the other hand, cannot be cured by orange light (Harris and Stark, 1977).

A relation between functionality of the phototransduction machinery and the integrity and maintenance of photoreceptor cells is suggested by the observations of milder forms of degeneration and morphological anomalies that develop slowly in the photoreceptors of *norpA* and *trp* mutants (Cosens and Perry, 1972; Alawi et al., 1972). There is also a depletion of rhodopsin with age in several severely mutant alleles of *norpA* (Ostroy, 1978). Indeed, Meyertholen and Ostroy, in unpublished work, have found that photoreceptors degenerate with age in these alleles, Thus, photoreceptors, like many other kinds of excitable cells, may require regular activity in order to be maintained (see later discussion on physiological activity in development). A process that may be relevant in this connection is an apparent turnover of membrane components induced by light observed in photoreceptors (White and Lord, 1975). By analogy with the work on bacterial chemotaxis, the study of interacting products of mutant genes is potentially a powerful way to understand a process such as phototransduction, particularly since the number of relevant genetic loci is likely to be small. Although the genetic map of *Drosophila* is, doubtless, not yet saturated with phototransduction mutants, a relatively large number of redundant alleles have been isolated for mutations defective in the photoresponse as measured by either the ERG or the PDA (Pak, 1979). The analysis of double mutants for interactions between gene products (such as the suppression of *rdgB*-induced degeneration by *norpA*) has not been applied extensively, but in one case it has helped to establish a link between a step in the PDA response and a step in phototransduction itself.

*Phototransduction Related to Prolonged Afterpotentials and*
*Biochemical Abnormalities*
Nine loci have been identified so far that, when mutant, cause flies to respond abnormally to stimuli that bring on the PDA, but normally to the standard white-light stimulus (Pak, 1979). All these mutants fail to depolarize for extended periods of time after being given a strong, blue flash. Interestingly, they fall into two distinct classes of PDA defect, one that exhibits the normal, prolonged inactivation of the photoresponse until an orange stimulus is given, even though it did not stay polarized (called inactivation-but-no-afterpotential, or *ina,* mutants), and another that showed no inactivation as well as no afterpotential (neither-inactivation-nor-afterpotential, or *nina,* mutants; figure 7). The relation of PDA responses in general to the actual phototransduction process was indicated initially by the finding of Minke and coworkers (1975a) that the PDA is composed of quantum bumps just as the receptor potential is. Preliminary genetic evidence for the interrelatedness of the two processes was provided by the interaction between a mutant of the *nina* type and the temperature-sensitive *norpA^{H52}* allele. S.K. Conrad, F. Wong, and W.L. Pak, unpublished, found that, in the double mutant, the *nina* mutation partially rescued the receptor potential that was otherwise abolished at higher temperatures in *norpA^{H52}* flies. It is not yet known whether this interaction is allele specific or whether mutants at more than one *nina* locus are capable of suppressing a *norpA* phenotype. However, further work on interactions between the effects of the *norpA* and *ninaA* genes may lead to important molecular information on how protein products of the loci associate during a critical early step in phototransduction. First, it has recently been suggested from some preliminary results that *ninaA* may possibly code for the protein component of a *Drosophila* rhodopsin (Pak et al., 1980). It is reasonable to speculate

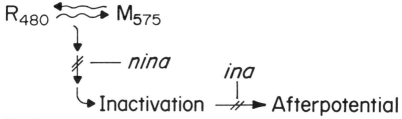

**Figure 7.**
Proposed scheme for the production of the prolonged depolarizing afterpotential (PDA) in *Drosophila*. The process is envisioned as consisting of at least two sequential steps: the inactivation step and the depolarization step. In the case of the *ina* mutants the depolarization step is blocked, leading to defects only in the afterpotential, while in the case of *nina* mutants the block occurs at the inactivation level, leading to defects in both inactivation and the afterpotential: $R_{480}$, rhodopsin absorbing maximally at $\sim 480$ nm; $M_{575}$, metarhodopsin absorbing maximally at $\sim 575$ nm. [Pak, 1979].

that rhodopsin and the product of the *norpA* gene interact in an early step of transduction, both as an a priori idea, given *norpA*'s proven role in the transduction process, and also because of the nearly normal phenotype of the *norpA-ninaA* double mutant.

Additional biochemical data that are of considerable heuristic value for further studies of these hypothetical protein-protein interactions are related to the emerging molecular investigations of *norpA*. Ostroy and Pak (1973) and Hotta (1979) have found protein differences in *norpA* versus wild-type adults. The latter study is promising because the differences were found only in the retinal layer of the eye and not in the lens layer, optic lobes, or other parts of the head. The differences from wild type had to do with two "spots" of stained protein seen after two-dimensional gel electrophoresis: these retina-specific polypeptides stained significantly more weakly with respect to flies expressing *norpA*. The same two spots were affected by six independently isolated *norpA* mutations (Hotta, 1979).

*Mosaic Analysis of Visual Mutants*
Another mutation that affects the photoreceptor cells themselves, but in a different way from those discussed above, is the tan (*t*) mutant, which has long been known to have defective vision (McEwen, 1918), as well as unusually light body color. More particularly, tan flies are missing the on- and off-transients in the ERG (Hotta and Benzer, 1969), which are due to synaptic activity in the optic lobes (Pak, 1975). This mutation does not affect the depolarization of the photoreceptor cells (Alawi and Pak, 1971), suggesting that tan has a primary defect in the optic lobes. Yet a mosaic study disproved this notion. This means that genetically mixed flies were constructed, each one part tan, part normal (see later sections for more details of these mosaic procedures). In these mosaics, every eye with mutant photoreceptors had an

ERG lacking the transients, and every genetically normal eye had a normal ERG (Hotta and Benzer, 1970).

In mosaics of the type used to study the action of tan, it is frequently the case that the eyes are of one genotype and the optic lobes of another (Kankel and Hall, 1976). Thus, in the mosaics of Hotta and Benzer (1970), there had to be several individuals with $t^+$ optic lobes, yet a mutant ERG (and vice versa). Apparently, then, the primary defect in this mutant is in the photoreceptors themselves, not in other tissues that might have been thought to be responsible. It could be that tan blocks communication between the site of photoreception and the synaptic terminal, interfering with the spread of a graded impulse that travels down the axon of the photoreceptor cell. Alternatively, the visual system in tan could have a specific defect in synaptic interaction between the receptor cells and the neurons in the optic lobes; if this is the site of action of the mutation, then the mosaic results demand that the defect be a presynaptic one.

Mosaics have been used to analyze other visual mutants, particularly *norpA* alleles that, like tan mutations, have been found to act autonomously in the retina (see review of Hall, 1978a).

There are many points of similarity between the process of phototransduction in *Drosophila* and chemotaxis in bacteria. The transduction pathway in bacteria has (1) receptors for the initial chemical signal, analogous to rhodopsin, (2) a series of intermediate steps, most of which are still "black boxes," and (3) a quantifiable output in the form of flagellar rotation, analogous to the photoreceptor potential. Thus, it is not surprising that similarities in the genetic analysis have emerged: the use of interacting genetic loci, alleles with partial activity, and saturation mutagenesis that likely has identified most or all of the genes controlling these pathways (since so many "repeat" alleles of previously discovered genes are found in subsequent searches for mutations). It can be hoped that genetic dissection of sensory transduction will be applied to a study of other kinds of stimuli and to additional organisms; some encouragement along these lines is provided by the suggestion that there is an auditory transduction mutant in mouse (Steel and Bock, 1980). The various genetic approaches for analyzing sensory phenomena may prove to be generally applicable strategies for studying pathway phenomena in neurobiology, as they proved to be so useful in the classical studies of biochemical pathways (for example, Beadle and Tatum, 1941).

## NERVE IMPULSES AND IONIC CHANNELS

Perhaps the most fundamental property of neurons, and one of the best studied, is their ability to generate impulses. This property, shared by certain protozoan cells as well as metazoan neurons and muscles, has been the subject of several neurogenetic studies. Since membrane excitability has been investigated for many years by the use of physiological techniques and has generated its own enormous literature, we shall concentrate primarily on how genetic variants are currently being used to augment the standard methods of analyzing the ion selectivity and voltage sensitivity of macromolecules in these membranes.

## Ionic Channel Mutants in Paramecium

*Paramecium* is the simplest organism used in the genetic analysis of excitable membranes. The information coming from experiments on a relatively few key genes has been concerned mostly with making explicit connections between behavioral and physiological phenomena and with the beginnings of an identification of membranous molecules that mediate action potentials. When excited, the membrane of the large *Paramecium* cell generates an action potential (figure 8; Eckert, 1972; Kung et al., 1975). These potentials are correlated with the behavioral excitation of these cells, in which various stimuli (chemical, electrical, thermal, or mechanical) cause the organism to respond by speeding up, slowing down, or backing up. These changes in swim-

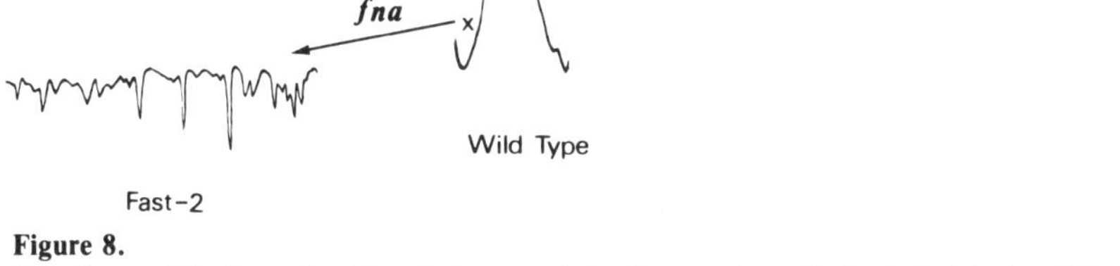

**Figure 8.**
Genetic modification of active electrogenesis in *Paramecium*. Variant electrical activities can be viewed as impairments of the wild-type activity by single point blockages resulting from genetic mutations. The center of the figure shows one episode typical of the recording from wild-type *Paramecium*, in which there is a long series of electrical responses to the appearance of 4mM NaCl, 1 mM Cacl$_2$, and 1 mM tris, *pH* 7.2, in the bath. In the case of the fast-2 (*fna*) mutant, blockage near the beginning of the depolarization activity by the *fna* mutation prevents depolarization completely. Blockage at the upstroke of the action potential by the *pw* mutation leads to the spikeless pattern of the pawn mutant. Blocking the downstroke of the action potential by the *Pa* mutation sustains the depolarization near the peak level for a much longer duration. Broken lines mark the estimated resting levels. Calibration is 10 mV, 1 sec. [Kung et al., 1975]

ming are governed by changes in the patterns of ciliary beating, involving the 5,000 cilia that cover each cell. Upon excitation, the cilia may change their frequency, direction, and amplitude of beating. How is the action potential mediated (for example, by which gene products), and how are the different possible changes in ciliary motion related to the ionic changes? These questions are strongly related to similar ones being asked about the mechanism of action potentials in higher organisms.

Also, the cilia of paramecia, like those found in many kinds of cells in metazoans, have a "dynein-tubulin" system of motility. In these higher organisms, cilia or modified cilia are known to be the sense organelles involved in the reception of stimuli such as odor and sound. In the future, then, certain features of motility and transduction in metazoans may be understood partially through work on paramecia (see discussion of Van Houten's (1977, 1978) work).

Approximately 300 behavioral mutants in *Paramecium aurelia* have been isolated (Kung, 1979), and a few additional ones in *P. caudatum* (Takahashi and Naitoh, 1978; Takahashi, 1979). The mutants were induced with a chemical mutagen and usually selected on the basis of defective behavior; some mutants were selected by nonbehavioral means, such as resistance to certain chemicals that interact with ionic channels. The latter mutations generally turned out to be in the same genes previously identified by behavioral mutants. In fact, the 300 mutations define only about twenty-five genes since many of the physiologically important genes have had multiple alleles defined for them, based on genetic mapping and complementation analysis. Parenthetically, it is worth noting that studies of these protozoa use many powerful genetic techniques developed by T.M. Sonneborn and students (see review by Preer, 1968). This background of genetic information goes, in fact, hand in hand with the physiological knowledge of these organisms, which dates back to the 1930s (see references in Eckert, 1972).

*Calcium Channel Mutants*
The genetic variants on which we shall focus most extensively appear to be defective in their voltage-sensitive calcium channels or in their potassium channels. The intracellular recordings from paramecia have shown that the action potentials that appear on stimulation of the cells involve a voltage-sensitive $Ca^{2+}$ conductance and a delayed rectifying $K^+$ conductance (Eckert, 1972). The mutations in genes called *pawnA, pawnB,* or *pawnC* were initially found (Kung et al., 1975) on the basis of having no avoiding reactions, for example, to sodium ions, which are repellents. Soon it was learned that these mutants can only swim

forward, with their cilia beating in only one direction. Moreover, the pawn cells from any of the three strains have no action potentials (figure 8), that is, no spikes in association with excitation though the membrane depolarization or generator potential seems normal. Thus, the *pw* mutants are candidates for having defects in the regenerative $Ca^{2+}$ channel proteins that mediate the inward current carried by this ion. It is possible that different components of such channels are specified by the different pawn genes. This assumes that a mutational alteration in any one of the supposed multiple components of the channel can knock out the entirety of channel function.

That the pawn mutants are defective in the ionic mechanisms that trigger reversal of ciliary beating, as opposed to a direct defect in the mode of moving the cilia, was demonstrated by experiments in which "models" of the paramecia were prepared. These cells are treated with a detergent that removes the membrane and thus all permeability barriers to ions. The models can be induced to swim forward (by adding $Mg^{2+}$, $K^+$, and ATP) and can show avoiding reactions in the presence of sufficient $Ca^{2+}$ (Kung et al., 1975). Models prepared from pawn strains can be induced to swim backward as well (Kung et al., 1975), which again suggests that the mutational defects involve the calcium channel and not the structure or function of the cilia. Parenthetically, we note here that proteins, especially mutant ones, can be thermolabile, and temperature-sensitive pawn alleles are not difficult to come by (Satow, Chang, and Kung, 1974). Such mutants exhibit heat-sensitive behavioral responses and action potentials, further suggesting that these genes code for channel proteins.

Work in the biochemical area has not progressed to the stage of identification and purification of "channel molecules," but there are tantalizing beginnings. First, it is known that the mechanisms for generation of action potentials are located in the membrane of the cilia, not in the part of the membrane covering the cell body. Thus, for example, cells denuded of cilia have no action potentials (Ogura and Takahashi, 1976; Dunlap, 1977). By stripping cilia off of the cells and preparing suspensions of these organelles, one achieves an immediate, manifold purification of the channel proteins. In fact, experiments on protein separation (gel electrophoresis and isoelectric focusing) have revealed some seventy different polypeptides associated with the cilia (Nelson and Kung, 1978). The putative "channel variants" have not yet been analyzed exhaustively; but differences in protein patterns between certain mutants and wild types have already been detected (cf. Adouette et al., 1980), although the origins of these differences are not yet very clear. Thus, we may anticipate that the calcium channel or other ion channels may be directly identified through the use of these mutants, and this may aid in our eventual understanding of the mediation of voltage sensitivity and ion selectivity at the molecular level.

Indeed, the calcium channel in this or any organism must be a complex molecular entity. Such an expectation is based on the long history or formal, physiological analysis of ionic channels from which we imagine that channels possess voltage-sensitive gates and also ion-selective pores (Ulbricht, 1977). Genetic evidence from *Paramecium* suggests that one may be able to extend this formal separation of channel components to a situation in which the components can be separated molecularly. For example, there are three different pawn genes that affect the calcium action potentials (Kung et al., 1975). A priori, the genes could all code for molecules with extremely similar functions. But the work of Schein (1976a,b; Schein, Bennett, and Katz, 1976) suggests that at least two of the pawn genes specify different components of the channel. Schein isolated several new alleles of the pawn genes, not by the behavioral selection used initially to discover these three genes, but, instead, by resistance to barium ions (which can carry the inward current, as can $Ca^{2+}$). These ions are poisonous to wild-type cells, and the strategy was based on the assumption that only those cells with altered calcium permeability would survive. The new mutants showed the anticipated decrease in excitability due to an alteration in calcium action potentials, just as did the previously isolated pawn alleles. An interesting pattern emerged from the careful examination of the several different mutants obtained in these genes: The array of *pwA* mutants had defects in activation of the inward calcium current, covering a range from moderate to severe decrements in the appearance of an inward current, while essentially all of the *pwB* mutants showed uniform abolition of the inward current. In addition, all the *pwB* mutants had a common permeability defect, manifested by a characteristic alteration in their anomalous rectification. From these results, Schein suggested that *pwA* may specify the gate molecule per se of the calcium channel. This hypothesis relies on the reasonable assumption that alterations of voltage sensitivity or time course of conductance changes could easily be expected to exhibit different degrees of mutational modification. There is, as yet, no direct evidence for this, as recent voltage clamp studies show no difference in the voltage sensitivity between the residual Ca conductance in a leaky *pwA* and the wild-type Ca conductance (Satow and Kung, 1980). It was the absence of any "leaky" mutants among the *pwB* group, coupled with the observed permeability defects, that suggested that this gene might code for the pore protein of the channel. Ion-selective aspects of channels suggest that there are great restraints on the size of pores. Thus, any mutational modification in a pore protein, even one that changes pore dimensions very slightly, might be expected to produce a very severe defect in channel functioning.

The ideas expressed here, though speculative, are of overt heuristic value for the molecular characterization of the putative channel proteins. Moreover, these facts and interpretations are a further indication of the value of isolating several mutant alleles of behaviorally important genes. On the other hand, deleting or changing the *function* of a channel does not necessarily imply deleting or changing the *structure* of the channel itself. The pawn gene products may not be the channel proteins, but they may function in the assembly, insertion, or maintenance of the channel, or in defining the immediate environment in which the channel opens and closes.

*Calcium Channel Mutants Used to Dissect Related Phenomena*
The pawn mutants have been used in several other ways to ask interesting questions about behavioral properties of *Paramecium* cells that do not necessarily bear on the molecular features of calcium channels. One can ask whether avoiding reactions are necessary in chemotaxis. It seemed possible that changes in the frequency of reorientation klinotaxis may underlie certain chemotactic responses. This was, in fact, shown when it was learned that pawn mutants, lacking klinotaxis, will not accumulate in higher concentrations of acetate or lactate, which are attractants for wild-type cells; taxes with respect to certain other chemicals are, however, normal.

Thermotaxis in paramecia also requires avoiding reactions, since pawn mutants, unlike wild types, will not disperse from heated regions (Hennessey and Nelson, 1979). For other stimulus-response systems, avoiding reactions do not seem essential, since pawn mutants are not noticeably different from wild types in geotaxis, galvanotaxis, and rheotaxis.

Apart from the contribution to the molecular dissection of channels, the pawn mutants have also been helpful in unraveling several other physiological problems. For instance, Oertel and coworkers (1977) were able to dissect components of the overall changes in membrane potential associated with the action potential. It was not known whether the calcium current is sustained or transient, irrespective of the repolarization effected by $K^+$ efflux. Using pawn mutants with no calcium current, Oertel and coworkers (1977) were able to subtract leakage currents and rectification currents measured in the mutant from the total current in wild-type cells (figure 9). Thus, the inward calcium current could be measured by itself, and it was, in fact, found to subside within 5 msec. In this experiment, the mutation was used analogously to a drug that would inactivate one channel and not the other. Because of nonspecific effects often associated with drugs, it is

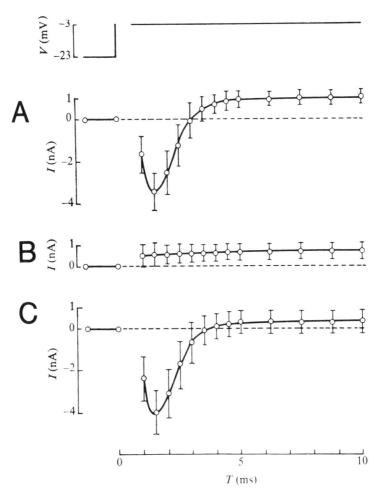

**Figure 9.**
Average time course of currents associated with step depolarizations of 20mV in wild type (*A*) and pawn (*B*). *Paramecium* mean current magnitudes ±S.D. were measured at various times from the same four wild-type cells and five pawn cells. (*C*) The difference between current measured in wild type (*A*) and (*B*) ±S.D. of the difference of the means. Measurements were not made in the first 1.0 msec after the step depolarization because the voltage was not well controlled during this time. [Oertel, Schein, and Kung, 1977]

sometimes more desirable to make these pharmacological manipulations with mutations. This point is also relevant in regard to neurochemical mutants in other organisms (see "Neurotransmission").

Since the depolarizing receptor potential, but not the spike of the action potential, is still present in pawn mutants (figure 8), one can now analyze the receptor potential in isolation and attempt to determine its ionic basis in an unobscured manner (Takahashi and Naitoh, 1978; Satow, 1981).

To estimate the lifetimes of calcium channels, pawn mutants have been used in association with special genetic methods available for *Paramecium.* Schein (1976b) studied pawn heterozygotes (*pw/* + ), which are essentially normal (the mutations are recessive). Pawn homozygotes (*pw/pw*) can be immediately produced, that is, in the next generation, by coaxing the cells into a self-mating process called autogamy (one of the key early discoveries of Sonneborn; (see Preer, 1968). The initial *pw/pw* cells still possess many normal calcium channels, though they and their subsequent offspring (from now on generated by the usual binary fission) cannot make active channels. In these experiments, active calcium conductance nearly halves in each fission, implying that the loss of channels is due simply to dilution and not to degradation.

Of the twenty or so behaviorally important genes other than *pawnA, pawnB,* and *pawnC,* few have been studied in enough detail to suggest explicit physiological defects. We should note that there are two additional pawn genes (D and E) that show the usual failure of avoidance, but only during phases of exponential growth; during the stationary phase, these cells are normal (Kung et al., 1975).

*Mutants Associated with Potassium Channel Function*
Three other genes for which mutants have been isolated may have to do with the potassium channel that helps mediate the action potential. Two of these are mutations, each in a separate gene, that confer a behavioral insensitivity to tetraethylammonium chloride (TEA) which is a blocker of potassium channels (Narahishi, 1974). The mutants were found on the basis of an absence of avoiding reaction to TEA; later it was demonstrated that they show a very weak avoidance to sodium (Kung et al., 1975). The first of these mutants isolated, *teaA,* has been studied physiologically and been found to have a large leakage of $K^+$ that "short circuits" the calcium activity mechanisms in the action potential (figure 10).

The other mutant that appears to be involved in potassium conductance, though not necessarily through the same channel affected by *teaA,* is Fast-2 (figure 8). These mutants, from any of more than thirty strains that have been isolated (all with mutations in the same gene),

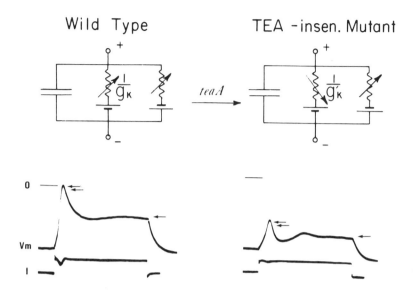

**Figure 10.**
Mutation affecting a $K^+$ channel in *Paramecium* membrane. The lower half of the figure shows the responses of the membrane of wild-type and the TEA-insensitive mutant to a $7 \times 10^{-10}$ amp, 20-msec injected outward current (*I* trace). The action potential (double arrows) on Vm trace) of the mutant is smaller than that of the wild type. The steady-state potential displacements (arrows) by the current are also different. Calibration is 10 msec, 10 mV; resting Vm, $-35$ mV for wild type and $-40$ mV for the mutant. Paramecia were bathed in 4 mM KCl, 1 mM $Ca^{2+}$, 1 mM citrate; 1.2 mM Tris, *pH* 7.2. The difference between the wild type and mutant membrane can be diagramed by the equivalent circuits shown in the top half of the figure, where the $K^+$ conductance is increased by the mutation. [Satow and Kung, 1976]

swim fast when disturbed in culture medium, but show no immediate avoidance of sodium. Avoiding reactions to TEA are normal. Measurements of relations between membrane potential level and concentrations of external ions indicate that Fast-2 is, in fact, defective in $K^+$ conductance. These experiments have revealed an abnormally large potassium permeability. There are no depolarizing reactions when sodium ions are applied; instead, the resting potential drifts toward hyperpolarization. TEA, when administered to these mutant cells, cures most of their behavioral and physiological defects (Kung et al, 1975).

The paranoiac mutant moves backward spontaneously for long periods of time. Again, there is inadequate information on the possible physiological basis for this behavioral abnormality. Yet there have been some interesting studies on the paranoiac mutants regarding their interactions with Fast-2 mutants and with respect to a putative ultrastructural defect. There are nine independent mutations that define four separate paranoiac genes (*PaA, B, C,* and *D*). All such variants not only show accentuated spontaneous avoiding reactions, but they also exhibit very prolonged backward swimming in sodium solutions. Physiologically, the mutant cells overreact to $Na^+$ in that there are prolonged depolarizations lasting up to 1 min (figure 8). The $Na^+$ influx is abnormally large, as is the $K^+$ efflux. The eventual repolarization of the membrane is much slower than that recorded from normal cells (Kung et al., 1975). A slow Ca-induced Na-inward current has recently been discovered in *Paramecium*. The *V-I* plot of this current shows a region of negative resistance and is therefore potentially capable of causing regenerative membrane depolarization. This Na current is much larger in a paranoiac mutant (Saimi and Kung, 1980).

Paranoiac mutants have been studied in double-mutant cells, that is, in the presence of Fast-2 mutations, which block the early depolarization events. The results show, predictably, that these two genes block different temporal stages of the action potential. Thus, membranes with both Fast-2 and paranoic-induced defects should, and do, show only the Fast-2 abnormality.

Recently, paranoiac mutants have been examined with respect to visible defects at the base of the cilia (Byrne and Byrne, 1978). Normally, one sees at the base an ordered matrix of membrane particles or plaques. One paranoiac strain, expressing a *PaA* mutation, showed many cilia (40 percent of those examined) lacking plaques altogether. Those with plaques had fewer, smaller, and poorly organized ones. Another allele of this *PaA* gene was also examined ultrastructurally and found to have a much milder abnormality in number and distribution of plaques; this allele is in the strain whose physiology has been examined most extensively. Thus, a fairly gross morphological abnormality is not a necessary cause or consequence of the severe behavioral and physiological defects. However, we are left with the reasonable suggestion that these membranous plaques are involved in $Na^+$ influx or $K^+$ efflux. This is an important area of future work, owing to its potential in establishing a connection between specific physiological events and particular features of ciliary movement that are crucial for the behavioral responses of this organism, and perhaps for the movement of cilia in general.

## Mutants Related to Ionic Permeability in *Drosophila*

### Action-Potential Mutants
Several mutations exist in *Drosophila* that could affect voltage-sensitive sodium channels. Any mutation in such channels would be expected to eliminate or alter the action potential in cells requiring

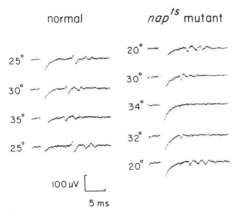

**Figure 11.**
Action potentials from wild-type and *nap^ts* *Drosophila* larvae. Signals from a small number of single nerve units excited by antidromic stimulation near the nerve terminals and extracellular recording from the segmental nerve bundle. In the *nap^ts* mutant, components dropped out individually as the temperature was raised and recovered when the temperature was lowered. Signals were averaged over 16 trials [Wu et al., 1978]

sodium for generation of impulses. Mutations in three different genes, all of which induce temperature-sensitive paralysis, cause the action potential in the abdominal motorneurons of the larva—as recorded extracellularly—to fail at the nonpermissive temperature characteristic of each mutant. Two of these mutants, no-action-potential (*nap^ts*; Wu et al., 1978) and paralyzed-temperature-sensitive (*para^ts*; Suzuki, Grigliatti, and Williamson, 1971), show sharp and reversible transitions for paralysis as adults and failure of neuromuscular transmission to the thoracic flight muscles at their restrictive temperatures (Siddiqi and Benzer, 1976; O. Siddiqi, cited in Pak and Pinto, 1976; Wu et al., 1978). Direct extracellular recordings from larval axons confirm that the nerve itself is defective in both *nap^ts* (figure 11) and *para^ts* (Wu and Ganetsky, 1980). Double mutants of the genotype *para^ts*, *nap^ts* fail to survive at any temperature (Wu and Ganetsky, 1980), suggesting that the two gene products may interact with each other. This evidence would be consistent with a putative lesion in the sodium channel if *nap^+* and *para^+* coded for different subunits. Recall the mutants of *Paramecium* (in which action potentials are mediated by calcium instead of sodium) and the three genes that, if any one has been mutated, are capable of altering or abolishing the function of the calcium channel.

More direct evidence for the involvement of at least the *nap* locus of *Drosophila* in regenerative sodium channels comes from experiments using the neurotoxins tetrodotoxin (TTX) and saxitoxin (STX), which are known to bind to and block the action of such channels in many organisms (Narahashi, 1974). Wu and Ganetsky (1980) have shown that larvae carrying *nap^ts*—under low-temperature conditions so that the gene product is not overtly turned off—exhibit action potentials with an increased sensitivity to the blocking effects of TTX. In perhaps a more direct assay of TTX binding influenced by this mutation, L.M. Kauvar, unpublished, showed that, compared to a homogenate from wild types, a particulate fraction from homogenates of *nap^ts* adult heads has altered binding properties, even when the assays (using radioactively labeled TTX) are done under mild temperature conditions. However, the details of Kauvar's results, and those of Wu and Ganetsky (1980) for that matter, are compatible with *nap^ts* reducing the *density* of sodium channels in nerve membranes, instead of the mutation leading to a change in the quality of individual channels. Indeed, L.M. Hall and J. Gitschier, unpublished, have obtained results suggesting that the former explanation may be correct, because the concentration of material that binds to STX (cf. Gitschier et al., 1980) is reduced in homogenates from *nap^ts* adults.

The apparent molecular complexity of ionic channels and their genetic control suggests that isolation of "channel mutants" by resistance to toxins may be an incomplete strategy. Such variants have been selected in vertebrate systems, for example, scorpion-toxin-resistant and batrachotoxin-resistant cells from a mouse neuroblastoma line (West and Catterall, 1979). But the use of poisons may not lead to the identification of all the channel components, even though the two toxins just mentioned apparently do bind to different "subunits" of the sodium channel, and TTX binds to yet another channel component (see references in West and Catterall, 1979). In any event, the isolation of physiological mutants and the construction of multiple mutant combinations of such variants will augment molecular-pharmacological approaches aimed at understanding the physical properties of these essential portions of neural membranes.

The third mutation in *Drosophila* that seems to cause temperature-sensitive failure of the action potential, comatose (*com*, Siddiqi, 1975; Siddiqi and Benzer, 1976), differs from the first two in requiring longer times for induction of, and recovery from, paralysis at high temperature, showing a graded decrease and recovery of synaptic function at the neuromuscular junction, and exhibiting cold sensitivity in addition to heat sensitivity for paralysis. The *com* variant survives as a double mutant with *nap^ts* (Wu et al., 1978). Recent intracellular recordings of action potentials (M. Tanouye, unpublished) cast doubt on the initital suggestion that *com* directly affects nerve impulses.

Recordings from the larval neuromuscular junction of *nap^ts* and flies expressing the other mutations that are thought, at least indirectly, to affect nerve impulses have shown an abrupt failure of transmission at temperatures above 35 °C, and both the muscle membrane and the release mechanisms in the synaptic terminal have been ruled out as sites of the genetic lesion (for example, Siddiqi and Benzer, 1976; Wu et al., 1978).

*Mutants Associated with Potassium Channels*
With respect to potassium-channel functioning, there are several interesting genetic and physiological findings, suggesting that a gene coding for at least a portion of this ionic channel has been identified. Most of the evidence concerns alleles of the Shaker locus in *Drosophila* (for example, Kaplan and Trout, 1969), which cause marked leg shaking, especially in etherized animals, and gross abnormalities in synaptic transmission at the larval neuromuscular junction (Jan, Jan, and Dennis, 1977).

Three different alleles of the Shaker locus were examined by Jan and coworkers (1977), all of which produced postsynaptic potentials in muscle that were longer and larger than in the wild type, particularly under conditions of low calcium concentration (figure 12). In addition, if calcium in the bathing medium was reduced to block transmitter release, and calcium was then applied focally from a micropipette near the nerve terminal (so that the timing of its availability could be controlled), they found that the Shaker mutants were still responsive to a calcium pulse long after the wild type had ceased to be responsive, as if the terminal were staying depolarized for too long. They provided evidence that the extended depolarization was not the consequence of overly active sodium channels by showing that even if the nerve terminal was depolarized focally, under conditions where the sodium channels were nonfunctional (that is, in the presence of TTX), the synaptic potentials characteristic of the mutants could still be elicited. They also tried to rule out a defect in the closing mechanism of the channels that permit calcium entry into the terminal (see Katz and Miledi, 1969; Llinás, Steinberg, and Walton, 1976) by showing that an abrupt hyperpolarization of the terminal was sufficient to arrest transmitter release. This suggested that the calcium channels were not "stuck" in the open configuration, but that they were merely responding to the time course of depolarization. Recently, direct evidence for a prolongation of the action potential was obtained by intracellular recording (following brain stimulation) from a giant axon in the neck of adults expressing this mutation (M. Tanouye and A. Ferrus, unpublished). These recordings also show abnormal shoulders in the action potential, after the initial spike, followed by multiple spikes (as opposed to the single spike always seen in these kinds of recordings from wild-type adults, after electrical stimulation of the brain; Tanouye and Ferrus, unpublished).

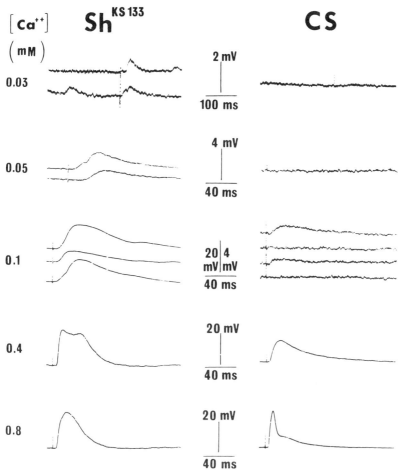

**Figure 12.**
Synaptic potentials recorded at different external [Ca$^{2+}$] from normal (CS) and mutant (*Sh$^{KS133}$*) *Drosophila* larvae. Stimulation artifacts are indicated by arrows. [Jan, Jan, and Dennis, 1977]

An observation leading to the suggestion that the potassium channels are the site of the defect was that two different drugs, 4-amino-pyridine (4-AP) and TEA, each of which blocks potassium channels in many organisms (Narahashi, 1974), mimicked the mutant phenotype when applied to wild-type neuromuscular junctions. Moreover, the resemblance was such that each drug specifically mimicked a different allele of Shaker. That is, the prolongation of depolarization (as inferred from the time during which synaptic release could be induced by a calcium pulse) was approximately 30 msec with 4-AP and 60 msec in *Sh$^{KS133}$*, while the duration after treatment with TEA or in the mutant

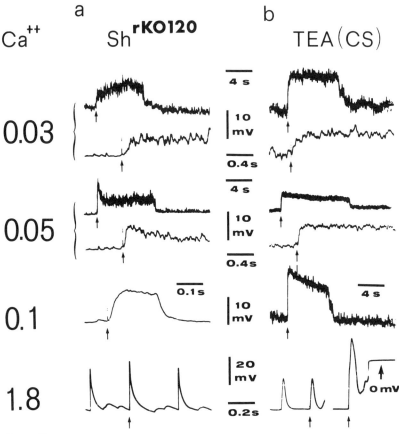

**Figure 13.**
Synaptic recordings at different external [Ca$^{2+}$] for (a) Sh$^{rKO120}$ larvae and (b) normal
(CS) *Drosophila* larvae treated with 10 mM TEA. The frequency of stimulation was one
per 30 sec. Arrows indicate stimulation artifact. [Jan, Jan, and Dennis, 1977]

Sh$^{rKO120}$ was two orders of magnitude greater (10 and 7 sec, respectively; figure 13). The behavioral phenotype of Shaker, leg shaking under etherization, could be elicited by feeding 4-AP to wild-type flies. A failure to mimic this behavior with TEA-fed flies was tentatively ascribed to a permeability barrier for TEA.

The possible correlation between two drugs known for their different effects on the potassium channels of the squid giant axon (for example, Armstrong and Binstock, 1965; Yeh et al., 1976), with two alleles of the same gene in *Drosophila* that exert analogous effects, is intriguing. The two drugs have been inferred to act at different sites on channel, and their effects are somewhat different. The block caused by 4-AP is less complete than that induced by TEA, it is partially reversible, and it requires greater depolarizations of the membrane (Armstrong and Binstock, 1965; Yeh et al., 1976). By analogy, the allelic mutations could conceivably represent lesions at different sites of a protein that makes up the channel. Conclusive attribution of the mutant defects to the potassium channel will require voltage-clamp measurements of the potassium currents across the membrane.

The existence of two alleles of the locus, with differing degrees of dysfunction, has again proved valuable in developing the hypothesis regarding the gene product of this locus, especially when comparing the effects of these two mutations with those of the drugs. However, it is a weakness of these findings and their implications that the putative potassium channel blockers available (such as aminopyridines) are not as well understood in their action (Yeh et al., 1976), as are, for instance, the sodium channel blockers discussed previously. In fact, there is no direct demonstration that these drugs actually bind to the potassium channels (Thesleff, 1980), thus making it all the more useful to have discovered the possible genetic disruptions of the potassium channel, since a drug-binding strategy for isolating molecular components of these particular channels is of dubious utility.

In this light, it is possible that strictly genetic and molecular-genetic experiments on *Drosophila* will be the approaches that will eventually result in the purification and molecular characterization of different ionic channels. For instance, Jan and coworkers (1977) and Wu and colleagues (1978) have suggested that the molecular cloning of genes in *Drosophila* could conceivably allow components of potassium channels and sodium channels to be synthesized in vitro. This would involve the identification of a cloned segment of *Drosophila* DNA, using already existing banks of such recombinant DNAs, in which many individual portions of the fruit fly genome have been inserted into prokaryotic vectors such as bacterial plasmids or bacteriophage. If such particular clones could be identified, and if further necessary advances in the experimental control of in vitro transcription and in vitro translation of eukaryotic genes should be forthcoming, then large amounts of, say, relatively pure potassium channel protein could be produced—much greater amounts than would seem to be extractable from homogenates of the flies themselves or, for that matter, from homogenates of a vertebrate brain.

In addition to the required advances in our understanding of molecular gene expression in vitro, this general strategy also demands a more sharply defined localization of the Shaker, *para^ts*, and *nap^ts* genes at their respective loci. It is relatively difficult to "clone a gene" for which no product is known. Thus, the locations of the just-mentioned genes, which are all in this category, may have to be pinned down to a particular chromosome "band" (regarding *Drosophila's* salivary gland chromosomes), using partially deleted chromosomes and other genetic aberrations, such as variants that happen to have a chromosomal "break" right in the gene of interest. Only then could a molecular clone putatively containing all or part of the correct sequences be correlated with the appropriate gene.

Certain genetic loci that code for, as yet, unknown products have, indeed, had certain of their parts molecularly cloned. These achievements, regarding the bithorax (Ashburner, 1980) and white (P. Bingham, R. Levis, and G.M. Rubin, unpublished) gene complexes in *Drosophila,* were greatly aided by the availability of chromosomal rearrangements having breaks in or very near the relevant loci.

With regard to the Shaker locus, recent experiments have nearly achieved the genetic prerequisites for the molecular cloning of the gene. However, these data (A. Ferrs and M. Tanouye, unpublished) also reveal some unanticipated complexities associated with this genetic locus. It has turned out that chromosomal translocations involving break points very near, or at, *Sh*'s location on the X-chromosome, known initially from its recombinational map position (Kaplan and Trout, 1969) do lead to a Shaker phenotype, both behaviorally and physiologically. Yet these break points can be in one of two *separate* locations, though they are only a very few chromosome bands apart. Furthermore, the small region between the two sites is special because it leads to *lethality* if only one copy of it is deleted from a fly that has two X-chromosomes (whereas in nearly all other regions of the genome, such small heterozygous deletions are compatible with good viability; Lindsley et al., 1972). This "haplo lethal" region is believed to be associated with potassium channel function, as are the two sites flanking it, because a small heterozygous deletion of the region has been found to allow survival if it is "covered" by a duplication of this segment deliberately placed elsewhere in the genome: A fly of this genotype shows the usual leg shaking and aberrant repolarization of its action potentials (A. Ferrs and M. Tanouye, unpublished).

The complexities of the Shaker locus are not worked out completely; but, instead of merely being puzzling, they might suggest that the separate sites within this apparent cluster of related genes (cf. "chemotaxis operons" in bacteria) specify different molecular components of a putatively multipartite potassium channel. Also, it should be considered that the various factors at the *Sh* locus may code for parts of different potassium channel *types,* of which there appear to be several (for example, sensitive to changes in voltage or to changes in internal calcium concentration; see review by Adams, Smith, and Thompson, 1980). In any event, the "broken" *Sh* genes may not only facilitate the molecular cloning of the DNA from this chromosomal location, but should also eventually lead to a molecular characterization of the different parts of the complex locus, which will, in turn, start to provide a complete understanding of the organization of these genetic sites and their possible relationship to polypeptides mediating potassium currents.

## Physiological Variants in Vertebrates

The information on genes controlling or contributing to vertebrate ionic channels is very limited, but the little that is known is significant. In addition to the aforementioned cellular variants of mouse (West and Catterall, 1979), the newt, *Taricha torosa,* has sodium channels that are resistant to the action of tetrodotoxin. These channels must be resistant since the newt embryo contains the toxin endogenously (Twitty, 1937). That the resistance is intrinsic to the nerve, as opposed to some general metabolic or compartmental protection, was shown in interspecific chimeras, in which neuronal tissues grafted from the axolotl *Ambystoma mexicanum* were affected by the toxin while growing side by side with unaffected *Taricha* neurons (Twitty, 1937).

## NEUROTRANSMISSION

The synthesis, release, reception, and degradation of neurotransmitters are obviously very important features of neurotransmission. A great deal is known about the function of the relevant molecules from nongenetic studies involving physiological, pharmacological, and molecular techniques. The purpose of augmenting these methods with the introduction of genetic variants is not necessarily to generate discoveries of new neurotransmitters, or even to prove that a particular molecule has such transmitter function in some organism; rather, mutants of neurotransmitter release, metabolism, or reception can be used to make perturbations of neurotransmission that are difficult to achieve otherwise.

## Mutants of Synaptic Action in *Drosophila*

### Facilitation
The modification of synaptic action by repeated stimulation of the synapse is of intrinsic physiological interest and has also been postulated to play a role in behaviors such as learning (for example, Kandel, 1976, 1978). Such a modification, that is, facilitation, has been found at the neuromuscular junction of *Drosophila* larvae, where repeated stimulation of the nerve results in an increase in the synaptic potentials. The study of a mutant with an abnormally high sensitivity to the facilitating effects of repetitive stimulation, due to the bang-sensitive

variant (Jan and Jan, 1978), has initiated genetic inquiry into the cellular mechanisms subserving plasticity at synapses.

The larval neuromuscular junction shows two kinds of facilitation. One operates over very short time intervals (for example, 20–200 msec); thus, for a pair of sequential stimuli, the response to the second stimulus increases 25–75% over the first, and shorter time intervals between stimuli lead to larger increases in response (Jan and Jan, 1978). An analogous phenomenon, studied at the frog neuromuscular junction (Katz and Miledi, 1968), has been ascribed to the persistence in the nerve terminal of residual calcium that entered during the initial stimulus. "Short-term" facilitation of this sort is normal in the bang-sensitive mutant.

A second kind of facilitation that has been studied in larva takes many stimuli to develop and can be made to last for minutes (Jan and Jan, 1978). Stimulation of the nerve at a rate of 10/sec for 20 sec or more results in a four- to fivefold increase in the size of synaptic potentials, an increase that can be maintained by continued stimulation at 4/sec. The mutant bang sensitive differs from the wild type in that the facilitation requires a lower frequency of stimulation (4/sec) for a shorter time (10 sec) in order to develop, and thereafter requires a lower rate of stimulation (1/sec) to be maintained. Thus, the mutant appears to be unusually sensitive to the factors that produce "long-term" facilitation and is abnormally slow to recover from it.

The finding of a mutant with defects in long-term facilitation, in the absence of abnormalities in short-term facilitation, defines a formal distinction between the two processes, consistent with evidence indicating different ionic requirements for the two phenomena (for example, Katz and Miledi, 1968; Atwood, Swenarchuk, and Gruenwald, 1975; Jan and Jan, 1978). Depolarization of the nerve terminal due to the internal accumulation of excess sodium has been postulated to underlie the increase in transmitter release characteristic of long-term facilitation (Atwood, Swenarchuk, and Gruenwald, 1975; Jan and Jan, 1978). Jan and Jan (1978) have shown that treatments that increase sodium accumulation in nerves by inhibiting the sodium-potassium ATPase can give a wild-type synapse the characteristics of a bang-sensitive synapse. These same treatments, in turn, make the neuromuscular junctions of bang-sensitive even more sensitive. The actual site of the mutant's defect has not been determined, and any explanation must be able to account for the apparently contradictory physiological defect observed in bang-sensitive: a twenty-fold decrease, relative to wild type, in the rate of spontaneous "miniature" potentials recorded at the synapse. The contradiction arises from the fact that the rate of miniatures tends to be *increased* by treatments that also increase the size of evoked synaptic potentials (for example,

Alnaes and Rahamimoff, 1975). Thus, a defective sodium pump would not seem to be sufficient to account for the complete phenotype associated with this bang-sensitive mutation.

*General Membrane Mutants*

Other temperature-sensitive mutations affecting synaptic transmission are in the shibire gene. *shi^{ts}* mutations, which will be mentioned again with regard to their derangement of development (abnormalities of cell differentiation in neuroblasts and myoblasts and faulty pattern formation of imaginal disk derivatives), were originally detected in tests of heat-induced paralysis (Grigliatti et al., 1973). Subsequently, the mutants have been found to have normal action potentials at high temperature; yet synaptic transmission at neuromuscular junctions (in larvae and adults) appears to fail when the temperature is raised (Ikeda, Ozawa, and Hagiwara, 1976; Siddiqi and Benzer, 1976; Salkoff and Kelly, 1978). Defects in potassium channels at nerve terminals have been implicated as being involved in the synaptic defects (Salkoff and Kelly, 1978). However, other abnormalities are also found at these terminals in *shi^{ts}*, such as depletion in the number of synaptic vesicles (Poodry and Edgar, 1979) and aberrant formation of membranes early in development (Swanson and Poodry, 1980).

The precise component of excitable cells coded for by shibire, which, when mutated, would be the site of the primary defect responsible for the behavioral abnormality and impairment of function at neuromuscular junctions, is not easily comprehensible. For instance, it would seem as if the shibire gene would not simply code for potassium channel proteins at neuromuscular junctions and still lead to the many developmental defects that it does when mutant (Poodry, Hall, and Suzuki, 1973; Buzin et al., 1978; Swanson and Poodry, 1980). The mutants are even rather pleiotropic in the physiological changes that they induce in adults, so that, in addition to the aforementioned neuromuscular defects, components of the ERG are altered in *shi^{ts}* flies (Kelly and Suzuki, 1974); there is a drastic increase in the firing of cells in several parts of the adult fly (Salkoff and Kelly, 1978; Koenig and Ikeda, 1980). There is even an anomalous transition in the Arrhenius plot related to the activity of a membrane-bound enzyme of mitochondria found in studies of *shi^{ts}* adults (and also in flies expressing several other temperature-sensitive paralytic mutations, including cold-sensitive variants; (Søndergaard, 1976, 1980). It may be that a generalized abnormality of membranes (see Swanson and Poodry, 1980) underlies the array of changes in cell development, orphology, and function induced by mutations of this gene. In spite of the vague and speculative nature of such a statement, it should be kept in mind that the function specified by the *shi^{ts}* may eventually turn out to be novel and critical for excitable cells. The puzzling nature of the pleiotropic effects of these mutations should not deter investigators from further study of the gene's action.

## Mutants Associated with Neurotransmitter Function

The enzymes of neurotransmitter metabolism are components of the nervous system that have been extensively studied and characterized in many organisms (Hall, Hildebrand, and Kravitz, 1974). They have been among the most accessible components of the nervous system to genetic study and manipulation owing to the ease with which they can be assayed and also to the powerful genetic technique of segmental aneuploidy (figure 14), which permits the structural locus for an enzyme to be found even if no genetic variants exist (O'Brien and MacIntyre, 1978).

Acetylcholine (ACh) appears to be a major transmitter in the central nervous system (CNS) of *Drosophila,* as in many other insects and invertebrates (Pitman, 1971; Gerschenfeld, 1973; Klemm, 1976). Mutations affecting the enzymes that are specifically present in the fly for metabolizing ACh have been found. Such an enzyme is acetylcholinesterase (AChE), which is distributed throughout the CNS (Hall and Kankel, 1976). *Drosophila* is also a rich source of the ACh-synthesizing enzyme cholineacetyltransferase (CAT; Driskell, Weber, and Roberts, 1978). In this particular insect, there is relatively little physiological information that definitely points to ACh as a central neurotransmitter owing to the technical difficulties in making recordings from the very small cells of the CNS. The absence of this information was, in fact, one reason for setting out to achieve mutational disruption of ACh metabolism.

To make a specific search for mutants of ACh metabolism, Hall and Kankel (1976) and Greenspan (1980) first discovered two separate chromosome segments that appeared to code for AChE and CAT, respectively. The genetic procedure (figure 14) involved the use of rearranged chromosomes to construct many different strains carrying three copies or one copy of individual chromosome segments in the genome (Lindsley et al., 1972). The normal number of copies of each segment is two. The flies with too much or too little genetic material are called aneuploids. Enzymatic assays on the partially aneuploid flies (for example, carrying three doses of a chromosome segment) showed changes in the levels of AChE activity (increased activity for three doses), exclusively for one small segment of chromosome *3;* another segment on this chromosome led to altered levels of CAT when the dosage of that segment was changed (figure 14).

Mutations in the AChE gene and CAT putatively contained in these two chromosome regions were induced by treating normal chromo-

## A. Gene localization by aneuploidy

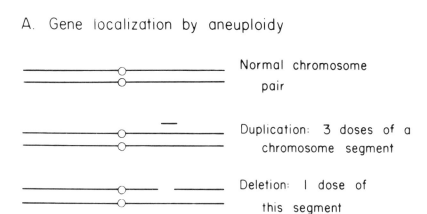

Normal chromosome
pair

Duplication: 3 doses of a
chromosome segment

Deletion: 1 dose of
this segment

## B. AChE gene localization

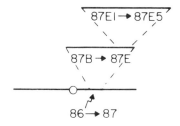

## C. CAT gene localization

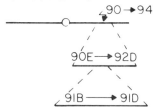

**Figure 14.**
Scheme for localization of genes for AChE and CAT by aneuploidy. (*A*) Doses of chromosome segments in duplicated or deleted flies. A series of duplications and deletions are constructed for nearly all the different subsegments of the four chromosomes of *D. melanogaster*. A schematic example of a duplication and a deletion of one such subsegment is shown. Duplication of a segment gives about 40% increase in activity of an enzyme whose structural locus is in the segment; deletion of a segment gives about 40% decrease. (*B*) Localization by dosage sensitivity of structural gene for AChE to successively smaller segments of the right arm of chromosome *3*. Numbers and letters labeling each segment refer to the standard map of *Drosophila* chromosome bands drawn from the polytene chromosomes in the larval salivary glands (Lindsley and Grell, 1968). (*C*) Localization by dosage sensitivity of structural gene for CAT to successively smaller segments of the right arm of chomosome *3*. Segments labeled as in *B*. [Hall, Greenspan, and Kankel, 1979]

somes with a chemical mutagen and then making these chromosomes heterozygous with other, third chromosomes that are *deleted* of the appropriate segment. It was thought that mutations in the AChE or CAT genes might lead to lethality. Some of the lethal mutations "uncovered" by the appropriate deletions (which are missing at least six genes or so) might therefore be in these two genes of interest. Subsequently, lethal mutations in both the gene for AChE (called *Ace*) and the gene for CAT (*Cha*) were isolated. For most of the newly induced mutations, lethality was first assessed at a relatively high temperature (29 °C), in hopes that some of the mutations would not cause lethality at lower temperatures (for example, 18 °C). Such temperature-sensitive alleles were, in fact, found and have allowed an extra degree of manipulation of AChE metabolism (Greenspan, 1980; Greenspan, Finn, and Hall, 1980).

The *Cha* and *Ace* loci are thought to represent the only genes for these enzymes in the fly (Hall, Greenspan, and Kankel, 1979), based on several pieces of evidence: (1) Only one dosage-sensitive region was found for either enzyme. (2) All detectable enzyme activity is absent from mutant embryos measured radiometrically for CAT and AChE as well as histochemically for AChE. (3) All detectable enzyme activity can be eliminated from adults bearing temperature-sensitive mutations of *Cha* or *Ace* by high-temperature treatments. (4) Nonconditional mutations of the *Cha* and *Ace* loci are lethal in the embryonic period and conditional mutants are potentially lethal in larval and adult stages as well (Hall and Kankel, 1976; Greenspan, Finn, and Hall, 1980; Greenspan, 1980; Hall et al., 1980a), indicating that there are not two or more genetic forms of either enzyme (see discussion of acetylcholinesterase mutants in the nematode).

Inferences as to the molecular characteristics of AChE, based on the *Ace* mutants in *Drosophila,* suggest that it is a multimeric enzyme, as is vertebrate AChE (Silman and Dudai, 1975; Bon, Vigny, and Massoulie, 1979). This is implied by the fact that temperature-sensitive mutants for *Ace* are virtually all cases of intragenic complementation. Further evidence for an interaction between subunits produced by different *Ace* alleles comes from the observation that adult flies heterozygous for the alleles $Ace^{j40}/+$ or $Ace^{lm15}/+$ exhibit temperature-sensitive paralysis and death (Greenspan, Finn, and Hall, 1980). This dominant phenotype suggests that the subunits contributed by the mutant alleles, even in a hybrid enzyme, are sufficient to destabilize it at high temperature. Two other Ace alleles that may prove interesting from the standpoint of the molecular characteristics of AChE produce detectable enzyme activity even though they are lethal (Greenspan, Finn, and Hall, 1980; Hall et al., 1980a). These mutations, one of which is cold sensitive, may affect parts of the enzyme other than the active site, such as sites responsible for assembly or sub-cellular localization. There is evidence that AChE in *Drosophila* homogenates is associated with a particulate fraction (Dudai, 1977a); thus, these mutants may be impaired in the enzyme's ability to associate with membranes. If CAT, which has been purified from *Drosophila* (Driskell, Weber, and Roberts, 1978), proves to have such a simple, monomeric structure, this will be consistent with the known lack of interactions between different mutant gene products (Greenspan, 1980).

These studies of ACh metabolism in *Drosophila* point out that systematic procedures are available for obtaining mutations in, and perturbing the function of, proteins of neurobiological interest. Wright, Hodgetts, and colleagues have also used the technique of segmental aneuploidy for discovering the gene in *Drosophila* that codes for the enzyme, dopa decarboxylase. The neurotransmitter dopamine has been detected in *Drosophila* (Tunnicliff, Rick, and Connolly, 1969), although no concrete neurological correlates of its presence have yet been found. Dopa decarboxylase, which performs the last step in the biosynthesis of dopamine in *Drosophila,* almost certainly participates in the process of cuticle sclerotization as well as in neurotransmitter metabolism (Lunan and Mitchell, 1969; Dewhurst et al., 1972). Wright and coworkers have identified the structural locus for dopa decarboxylase by segmental aneuploidy and obtained mutations in this gene, designated *Ddc* (Hodgetts, 1975; Wright, Bewley, and Sherald, 1976; Wright, 1977). By showing that dissected ganglia from prepupae heterozygous for the mutant alleles (that is, $Ddc^- / +$ ) have 50% of the enzyme activity found in homogenates of wild-type nervous system tissue, Wright (1977) determined that the *Ddc* locus is responsible for the nervous system enzyme as well as for the epidermal one. Some of the conditional mutants at the *Ddc* locus (Wright and Bentley, 1979) are cases of temperature-sensitive intracistronic complementation (T.R.F. Wright, unpublished). $Ddc^-$ mutations are homozygous lethals, and they may have this severe effect, in part, because dopamine synthesis in the CNS is curtailed. Other cases of genetic abnormalities of catecholamine metabolism are dealt with in a later section on abnormal higher behavior in mammals.

L.M. Hall and colleagues have used segmental aneuploidy to search for genes in *Drosophila* that affect a putative ACh receptor, that is, a component that binds the snake venom, $\alpha$-bungarotoxin ( $\alpha$-BT). The binding assays used to identify this component of the fly's CNS have been performed on homogenates and also in sectioned material. These experiments show that this activity is present in nearly all synaptic areas of the CNS (Dudai and Amsterdam, 1977; Schmidt-Nielsen et al., 1977; reviewed by Dudai, 1979; Satelle, Hall, and Hildebrand,

1980), and one such putative receptor has been purified (L.M. Hall, 1980; Rudloff, Jimenez, and Bartels, 1980).

No genetic loci that might specify parts of the $\alpha$-BT receptors were found in the segmental aneuploidy screen (L.M. Hall, 1980). However, additional approaches have been used to achieve preliminary genetic information on ACh receptors. Hall and coworkers (1978) screened various wild-type strains for resistance to the insecticide and cholinergic ligand, nicotine, which has been used to characterize one of the receptors in *Drosophila* (Schmidt-Nielsen et al., 1977; Dudai, 1978). Eight resistant strains were found, and subsequent tests for alterations in their ACh receptors revealed that three of the strains had receptors with variant isoelectric points (Hall et al., 1978). Preliminary genetic experiments with one of these strains, isolated from a wild-type *Hikone-R* stock, have indicated that both the isoelectric variation and the major resistance to nicotine segregated with the X-chromosome (L. Hall, 1980).

These preliminary data on a gene affecting ACh reception in *Drosophila* may lead to the isolation of several mutant alleles of the X-chromosomal locus putatively coding for at least a portion of the nicotinic ACh receptor. It should be stressed that the purpose of such mutant isolation would *not* be solely to aid in purification and characterization of ACh receptor subunits. Such biochemical tasks can rather readily be accomplished by using toxins that bind to these receptors, especially since binding is very tight with respect to ACh receptors, in particular, and poisons such as $\alpha$-bungarotoxin or cobra toxin (see reviews of Fambrough, 1979; Morley, Kemp, and Salvaterra, 1979). Hence, more important reasons for inducing further ACh receptor mutants—some of which will no doubt be lethal—will involve manipulation of CNS development and function, as discussed below with regard to the other genetic variants of ACh-mediated neurotransmission.

What neural defects have been revealed with the mutants, associated with ACh action, that do involve variants sufficiently severe to be useful experimentally? In the current discussion, we shall focus on the physiological and neurological defects per se. In later sections we shall discuss these neurotransmitter variants as they bear on behavioral questions and also developmental ones relating to morphological abnormalities and lethal effects of *Ace* and *Cha* mutations. In fact, it is not easy to use these mutants to study physiological abnormalities because the animals expressing these lethal genotypes die at the end of the embryonic period, before much, if any, of the functioning of the CNS has begun. However, it has been possible to create genetic mosaics that are missing the normal gene for *Ace* or the normal *Cha* gene in only portions of the nervous system. The techniques

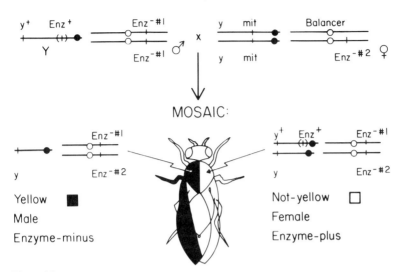

**Figure 15.**
Scheme for generating genetic mosaics for enzyme mutations. The male parent donates an X-chromosome (centromere = solid circle) carrying an insertion of the wild-type enzyme locus ($Enz^+$) and the wild-type allele of the cuticle marker $y^+$ (not-yellow); and an autosome (centromere = open circle) carrying a mutant allele of the enzyme locus ($Enz^{-\#1}$). The female parent donates an X-chromosome carrying the mutant allele of the cuticle marker $y$ (yellow) and an autosome carrying another mutant allele of the same enzyme locus ($Enz^{-\#2}$). Since the mother is homozygous for the mutation *mit* (mitotic-loss inducer) on her X-chromosomes, her progeny will have mitotically unstable chromosomes. Thus, a zygote that starts out as diplo-X can undergo loss of one of its X-chromosomes. If the father's $Enz^+$-bearing X is lost, it will be detectable by the fact that the recessive cuticle marker $y$ will be uncovered. Patches of yellow cuticle on adults will be diagnostic of mosaicism in that animal, and some of these mosaics will have lost the X with $Enz^+$ in their nervous systems as well. [Hall, Greenspan, and Kankel, 1979]

for generating these particular mosaics involve somatic loss of an X-chromosome that carries the normal gene of interest; the remaining alleles of this gene in such tissues will be mutant, and thus their effects will be uncovered. Other tissues in the same individual will not have lost this X-chromosome and so will retain the normal enzymatic function specified by the gene. Some of these neural mosaics, then, may survive, since they express the lethal factor in only part of their CNS. This kind of chromosome loss mosaicism in *Drosophila* has been used many times in neurogenetic analysis (Hall, 1978a; Palka, 1979a). A rather general genetic scheme used to make these mosaics is shown in figure 15. Other analogous methods are explained in a later section on genetic fate mapping.

The mosaics in *Drosophila,* produced by genetic techniques of induced chromosome loss, have mutant clones of tissue in essentially any part of the animal, including any part of the nervous system. Thus, the distribution of mutant and normal tissue seen among many

Physiological and Neurochemical Genetics

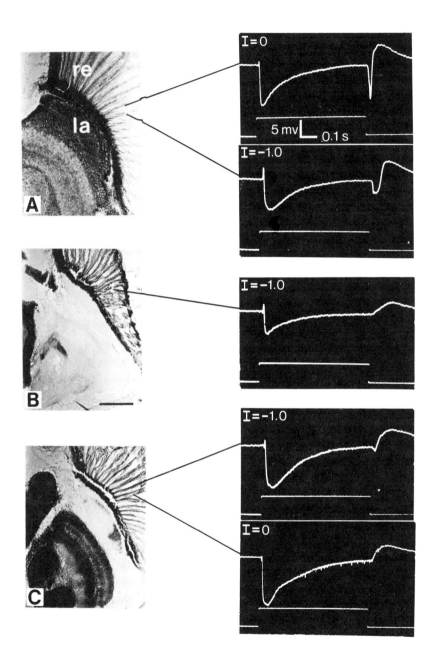

mosaics is in no way fixed from case to case. This situation is desirable, because one would like to know about altered cholinergic function in many different parts of the CNS.

The physiological findings on *Ace* mosaics were obtained from recordings of the visual system. It has been found that the electro-retinogram (ERG) is missing only its light-off-transient, if the first-order optic lobe in the visual system is AChE-minus (figure 16). The photoreceptor depolarization and on-transient are normal, implying that the cells of the eye function normally in these mosaics (as expected) and that there is partially normal synaptic activity in association with these mutant clones in the visual system. This notion is based, in part, on the demonstration that the on- and off-transients reflect synaptic activity in the optic lobes (see, for example, Pak, 1975). Nearly all mutants with defects in the transients but normal photoreceptor depolarizations (Pak, 1975) are missing both transients and are behaviorally blind or near blind (although see the recent description of ERG defects in the stoned-temperature-sensitive mu-tant in Kelly, 1981). But a defect in only the off-transient in the *Ace* mosaics does not have the same behavioral consequences. Specific im-pairments of visually triggered behavior are found in association with AChE-minus tissue in the optic lobes and in other nearby portions of the CNS (see the discussion of these findings in "Behavioral Neuro-genetics").

The physiological abnormalities, and the concomitant behavioral ones, could be directly due to faulty hydrolysis of ACh. Or perhaps these physiological defects are only secondary consequences of the absence of enzyme. This possibility is raised because there are

**Figure 16.**
Electroretinograms from normal and AChE mosaics in *Drosophila*. Photographs on the left sides how horizontal sections through the compound eye and optic lobes of each fly tested. The sections were stained for AChE activity. The retina (re) in each eye shows the regular array of ommatidia containing the eye-screening pigments. The optic lobes in *A,* including the first-order optic lobe, the lamina (lm), stain normally for AChE. The mosaic shown in *B* had no detectable AChE activity in its lamina and virtually none in the more central lobes. The mosaic shown in *C* had mutant tissue in most of the lamina, but not in the more central optic lobes. A small patch of AChE staining can be seen in the posterior lamina of this fly. The scale bar in *B* is 50 $\mu$m and applies to *A* and *C* as well. Electroretinogram traces for each fly are shown in the photographs on the right. The top trace in each photo is the voltage difference recorded between an elec-trode placed just under the cornea at the level of the retina and a reference electrode placed in the hemolymph of the abdomen, displayed on an AC-coupled oscilloscope. The bottom trace in each photo indicates the onset, duration, and end of the white-light stimulus. *I* values refer to log attenuation by neutral density filters of the full intensity (that is, $I = 0$), where full intensity corresponds to about $4 \times 10^{-3}$ W/cm$^2$ measured at 470 nm. The voltage and time scales, indicated in the second trace of *A,* are the same for all traces. [Greenspan, Finn, and Hall, 1980]

morphological defects induced in the CNS by the *Ace* genotype, and they may be the main cause of altered optomotor responses or court-ship behavior. The temperature-sensitive *Ace* mutations will be useful for addressing these questions because one could raise *Ace* mosaics carrying such alleles at the "permissive" (low) temperature and then raise the temperature for a brief period of time. These flies might then show the usual physiological and behavioral abnormalities in the absence of the structural defects, which probably could not be induced by only short-term heat treatment of these adults.

A final question raised by the work on *Ace* mosaics is; What are the neurochemical consequences of the enzymatic deficits? An absence of AChE might be expected to result in elevated levels of ACh (for exam-ple, Katz and Thesleff, 1957a). This could potentiate postsynaptic responses in mutant clones, but perhaps only transiently, if it is the case that the target cells and their receptors become desensitized. Arti-ficially accentuated levels of neurotransmitters have sometimes been found to cause structural damage (Landauer, 1975; Olney and de Gubareff, 1978; Wecker, Laskowski, and Dettbarn, 1978; Leonard and Salpeter, 1979) and also to result in desensitization (Tauc and Bruner, 1963; Katz and Thesleff, 1957b). The specific question, then, is, Could *no hydrolysis* of ACh in a part of the fly's CNS have the same effect as *no ACh* at all? Again, we can turn to genetic techniques for a possible answer to this question.

Mutant clones of CAT mosaics, analogous to the *Ace* mosaics, can be assessed for morphological defects that might be similar to those in *Ace* mosaics. One piece of physiological information on *Cha* mutants has already shown a result somewhat similar to that seen in *Ace* mosaics. *Cha-ts* flies were raised at their permissive temperature (18-22 °C) and then placed at 30-32 °C as adults. When these flies had passed out but were still alive (which requires 2-3 days of heat treatment, though wild-type adults do not pass out at these elevated temperatures), their ERGs were recorded. As in the *Ace* mosaics with mutant clones in the visual system, the *Cha* mutants expressed dramatic defects in the off-transient, whereas the on-transient and receptor potentials were normal (Greenspan, 1980). Therefore, it appears that the same synaptic component of the ERG is impaired by either kind of mutationally altered ACh metabolism.

**Neurochemical Mutants in Nematodes**

*Catecholamine Mutants*
Several types of mutants have been isolated in the nematode that have defects in various aspects of the function of the presumptive transmit-ter substances (reviewed by Riddle, 1978). The early work on genetic

aspects of neurotransmission in this organism concerns dopamine-containing cells that are probably involved in mechanoreception. These cells, of which there are eight in the hermaphroditic form of *Caenorhabditis elegans* and fourteen in the male of this species, have been detected by formaldehyde-induced fluorescence (FIF). Induced mutations affecting dopamine FIF were isolated by Sulston and coworkers (1975) by simply screening putative variants for abnormal fluorescence. Several mutants were found, defining six separate genes. The different mutant genes fell into two classes: the first affects the putative neurotransmitter in these cells, and the second affects the cells themselves. In the first class are mutants with reduced or undetectable fluorescence (*cat-2* and *cat-4* mutants, where *cat* stands for catecholamine-defective) or reduced fluorescence in the cell bodies of these cells but not in the cell processes (*cat-1* and *cat-3* mutants; *cat-1* mutants lack serotonin in their processes as well as dopamine). In the second class are two other mutants that affect development of the cells, as opposed to the neurotransmitter within them (for example, mutants with misplaced cell bodies or other mutants with over-production of dopaminergic cells; see discussion of cell lineage of the nervous system in "Developmental Neurogenetics").

Some behavioral tests have been performed on these "dopamine-defective" mutants, such as touch sensitivity measurements; yet even the mutants that are apparently deficient in a neurochemical sense are behaviorally normal (Sulston, Dew, and Brenner, 1975). Even double mutants of *cat-1* and *cat-2* responded normally to touch. These double mutants were used in further mutagenesis work, in accordance with the idea that more than one class of mechanoreceptor (for example, those involving and those not involving catecholamines) must be defective to give a behavioral abnormality. Indeed, eight new mutants relatively insensitive to touch were found; but all were shown later to be defective by themselves in $cat-1^+$ and $cat-2^+$ backgrounds.

Additional information on these two genes concerns the development of the anterior nerve ring and ventral ganglion in the worm. With either kind of mutational defect, no morphological differences from wild-type were found by using the ultrastructural reconstruction techniques that have been developed for the nematode, and carried to completion in the laboratories of S. Brenner and R. Russell (see references in the review of Riddle, 1977).

*Mutants of Acetylcholine Metabolism*
Unlike catecholamine mutants, which lack neurological abnormalities, mutants of ACh metabolism in *C. elegans* have already exhibited such neurological problems. Pharmacological work first indicated that ACh induces muscle contraction in the worm and that

the response is specifically due to nicotinic ACh receptors (Del Castillo, De Mello, and Morales, 1963). Cholineacetyltransferase is limited to a set of excitatory interneurons in the related roundworm *Ascaris* (Stretton et al., 1978), which may have a neuromuscular system homologous to that in *C. elegans*. Thus, ACh is likely to mediate excitatory synaptic functions in some neuromuscular junctions in these roundworms.

Russell and colleagues (Russell et al., 1977; Johnson et al., 1981) have searched for mutations affecting AChE in the nematode. Normally, the worm has five molecular forms of AChE, two of which are relatively sensitive, and three of which are essentially insensitive to inactivation by the detergent desoxycholate. Approximately 200 nematode mutants, nearly all of which were isolated on the basis of generally uncoordinated movements, were assayed for AChE. One mutant strain was found that lacked detergent-resistant activity, and it was soon shown to lack the appropriate three peaks of AChE.

The initial findings on this mutant point out the extreme importance of careful *genetic* characterization of putative neurochemical-behavioral mutants. The uncoordinated phenotype and the absence of some of the AChE in this organism turned out to be coincidental, not causal. Genetic mapping, carried out to extensive resolution by Johnson and coworkers (1981), showed that the mutations responsible for the behavioral and enzymatic defects are linked closely together on the same chromosome, but are in distinct genes. A corollary to this fact is that the enzymatically defective mutant has no detectable behavioral abnormalities whatsoever, and the behavioral mutant has a completely unknown biochemical etiology and normal AChE (Johnson et al., 1981). In the absence of this detailed genetic mapping, one might have assumed that the two phenomena were related, and proceeded with experiments based on this false assumption. Fortunately, this possibility was avoided, and we can now consider that both forms of AChE in the nematode have essentially the same tissue distribution (so that an absence of one form is compensated for by the other). This notion is given some support by recent work of Culotti and coworkers (1981), who mutagenized a strain with the original AChE mutation, now called *ace-1,* but free of the uncoordinated mutant. Indeed, a new strain was isolated that is missing approximately 95% of the total AChE in the organism. This strain proved to carry two mutations, *ace-1* and *ace-2,* which are unlinked. *ace-2* thus affects the detergent-sensitive form of AChE and, by itself, leads to no noticeable behavioral problems. The double mutant, though, is slow moving, uncoordinated, and hypercontracts when tapped. These abnormalities are what would be expected from treating the nematode with ACh agonists or anti-AChE drugs. Culotti and coworkers (1981) have found that the enzymatic defect (that is, nearly no AChE) and the behavioral defects cosegregate. Thus, it is now rather likely that the behavioral abnormalities are truly caused by the severe decrement in AChE. Also, it is possible that all of the AChE in the organism is coded for by these two genes; additional more defective alleles of *ace-1* and *ace-2* would be completely devoid of this enzyme activity, and such doubly mutant genotypes might, in fact, be lethal.

As in the case of the *Ace* mutants, the initial isolation of putative CAT mutations (in a gene called *cha-1*) in the nematode was somewhat indirect. One mutant was isolated according to the criteria of resistance to trichlorfon (a putative agonist of ACh receptors) and was rather paralyzed and uncoordinated. Extracts of this mutant were shown to have only 2% of the wild-type level of CAT (Russell et al., 1977). Are these enzymatic and motor defects directly connected? The lesson from the AChE story obviously must be applied here as well as to another mutant in nematode affecting CAT. This second strain was found on the basis of defective defecation behavior, but was otherwise apparently normal in movements (Russell et al., 1977) and subsequently proved to have less than 5% of the normal CAT levels (Russell et al., 1977). It is possible that the difference between a 95% and a 98% reduction in the activity of this enzyme is biologically significant, but much additional work is necessary.

Genetic dissection of the ACh receptor may prove very useful for studying the importance of cholinergic function in the nematode. To begin such experiments, Lewis and coworkers (1980) have recently isolated many mutants resistant to levamisole, an ACh agonist. The mutants with high resistance define seven different genes, and all exhibit uncoordinated movements. The mutations in these genes also fail to respond to cholinergic agonists that are effective in causing contractions of the wild-type (by using a fairly gross behavioral assay on cut-open worms, to which pharmacological agents can be applied). Some of the alleles of five of these seven genes lead to milder abnormalities, that is, intermediate resistance to levamisole and only a partially defective response to drugs. It is not clear whether all these genes directly code for components of ACh receptors or even directly affect receptor function. That ten separate genes can be involved in the response to an ACh agonist suggests that the involvement of some of these loci in cholinergic function is nonspecific.

One feature of the genetic approach toward studying neurotransmission in the nematode holds much promise for future work. Since the neuromuscular anatomy and cell lineage of this organism are so well worked out, and since the organism appears to be so simple (possessing only about 300 cells in its nervous system), an examination of a possible role for signaling molecules in cell differentiation and pattern formation could make powerful use of neurotransmitter mutants. For instance, is there a lineage relation among cells that contain a particular neurotransmitter? Would mutations that impair the synthesis of a neurotransmitter (for example, the current CAT variants) end up being behaviorally defective because the connections involving cells in this lineage are abnormal? Might the cell division in the generation of the lineage be abnormal? Also, will excitable cells that are targets of the transmitter-defective cells in this lineage fail to differentiate properly? (Compare cases of target cell differentiation in the vertebrate superior cervical ganglion when it is deprived of input from ACh-containing neurons; see for example, Black and Geen, 1974.) It should be noted here that it is difficult to obtain the nongenetic background information that is necessary in order to ask the above kind of questions about the nematode. Staining for the presence of a putative transmitter (for example, dopamine) can be done, but physiological characterization and manipulation of transmitters at specific synapses in this tiny animal are very difficult. However, it appears as if physiological data on *Ascaris,* which are beginning to be collected, may be applicable to the nematode. The *Ascaris* findings include information on which synapses in particular ganglia are excitatory and which are inhibitory (Stretton et al., 1978). No genetic manipulation is possible in *Ascaris,* but there are many apparent homologies in the morphology and cell lineages involving excitable cells in the larger worm and in the nematode. Thus, one may be able to connect genetic results from manipulation of the lineages and the transmitter functions in these cells with neurochemical data from the more physiologically accessible organism.

## Levels of Neurochemically Important Molecules
## Measured in Mouse Brain Mutants

The distribution of neurotransmitters in mouse brain has been studied in several neurological mutants (see review of Mallet, 1980). In particular, the occurrence of a given transmitter in association with a specific cell type has been assayed in mutants that eliminate that cell class. For example, the single-gene mutation responsible for Purkinje-cell degeneration (*pcd*) leads, postnatally, to a selective loss of essentially all of the cerebellar Purkinje cells. Such cells are thought to release the inhibitory neurotransmitter $\gamma$-aminobutyric acid (GABA) Indeed, GABA levels are much reduced in the mutants after degeneration has occurred (Roffler-Tarlov et al., 1979).

Another transmitter, glutamate, is very likely not released by Purkinje cells, and levels of this amino acid are normal in cerebella from homozygous nervous (*nr*) mice, which also lack Purkinje cells

(Roffler-Tarlov and Sidman, 1978). Glutamate is, however, believed to be associated with cerebellar granule cells, and the mutations stag-gerer (*sg*) and weaver (*wv*)—which specifically lead to a decrement in granule cells—exhibit reduced levels of this substance in the appro-priate portions of their mutant brains (Roffler-Tarlov and Sidman, 1978). These results are complicated by the fact that the deep nuclei of the cerebellum in *sg* and *wv* mutants also have a reduced amount of glutamate, even though there are no granule cells in this region of the cerebellum.

The difficulty in interpreting these findings brings out one of the limitations inherent in the use of mutants in which apparently only one specific class of cells has been eliminated. Pleiotropic effects of the mutations acting directly on other cell types (though not neces-sarily causing a blatant morphological change in these cells) or the absence of outside influences normally mediated by cell interactions from the missing neurons can potentially produce phenotypic changes at least as gross as the actual loss of the one cell type. Therefore, results of this sort must be handled with caution and analyzed more thoroughly than by simply measuring an isolated characteristic of a mutant (see below for a discussion of inductive cell interactions in "Developmental Neurogenetics").

Mutant *nr* brains have recently been analyzed with respect to their binding activity for the anticonvulsant benzodiazepine and there is a marked reduction in binding activity (Skolnick et al., 1979). This result suggests that these kinds of receptors may be confined to Purkinje cells. Another mutant, jimpy, was also used to probe binding activity related to opiate action in the mouse brain. The jimpy muta-tion leads to a severe decrement of myelin in the CNS and also to a deficiency in brain sulfatides. It has been suggested that cerebroside sulfate may be an important opiate receptor in nerve cell membranes (for example, Loh et al., 1975). To test this further, Law and coworkers (1978) measured both morphine sensitivity (by a behavioral test) and binding activity to this drug (by in vitro assays using synap-tosomal plasma membranes) in jimpy mice. Both the sensitivity to, and the number of binding sites for, morphine were reduced in the mutant, reinforcing the idea that this membrane acidic lipid has a role in binding morphine and mediating its effects.

# 3
# BEHAVIORAL NEUROGENETICS

## MOTOR MUTANTS

### Motor Output in Bacterial Chemotaxis

The later steps in the pathway controlling bacterial responses to chemical stimuli, introduced in the section on "Sensory Mechanisms," have been dissected by mutants that are generally defective in these behaviors. Such mutations lead to abnormal responses to all kinds of chemicals, not just certain classes of compounds (see figure 1). Two genes in this pleiotropically abnormal category (in the *che* category, first discovered in *Salmonella*) are now known to specify functions of methylation and demethylation of the methyl-accepting chemotaxis proteins (MCPs) (see "Sensory Mechanisms"). The other *che* genes have, as yet, unknown biochemical functions, but one can make hypotheses about the actions of these genes with respect to the linking of reception and signaling of chemical information to the motor output related to smooth swimming and tumbling.

Bacteria, such as *E. coli* and *S. typhimurium,* swim by rotating their flagella, which cover the surface of the cell and bundle together into a filament. This filament is essentially a helical propeller or corkscrew. Smooth swimming is generally caused by *counterclockwise* rotation of the flagella, allowing the rotating bundle to stay together and propel the cell forward. Tumbling is usually associated with *clockwise* rotation, which causes the bundle to fly apart and swimming to cease. Mere observation of swimming cells and their flagella does not lead to the conclusions just noted. Such observations might have led to the notion that the flagella propagate a wave rather than rotate (for example, Calladine, 1974). These possibilities were distinguished by the use of genetic and immunological tools. Silverman and Simon (1974) used double mutant *E. coli,* which has a "polyhook" at the base of the flagella (instead of the normal single hook) and has no flagellar filament. Such mutant cells are nonmotile, as expected; but the addition of antibody against the polyhook to a suspension of cells results in their attachement to microscopic slides at their hooks and thus to rotate rapidly! Other cells will attach to one another at their hooks and counterrotate. Following this discovery, Adler and colleagues (Larsen et al., 1974) showed that the addition of attractants accentuates counterclockwise rotation, in the above kind of behavioral assay, whereas repellents cause clockwise rotation. These important results make an explicit connection between chemical stimuli and the particular feature of motor output that mediates the organism's response to such stimuli. Moreover, it reinforces the fact that temporal, but not spatial, changes in the levels of these chemicals are critical (cf. Macnab and Koshland, 1972), since the tethered cells in the behavioral assay responded to their stimuli in precisely predictable ways.

How are changes in levels of the methylated proteins communicated to the machinery at the base of the flagella that rotate the filaments? There are several good hypotheses to be made from the studies of the other *che* genes and certain of the genes involved in motility per se. The detailed experiments on genetic mapping and complementation have shown many of the *che* and motility genes to be very closely linked on the bacterial chromosome. Some of these genes are even organized into an operon, a group of adjacent genes with related function. The close physical association of these genes implies that their expression is coordinately regulated (figure 17; Silverman and Simon, 1974; Warrick, Taylor, and Koshland, 1977). Thus, we have an example of a fundamental molecular genetic concept, coordinate regulation, pertaining to genes whose cellular functions contribute to the unique character of the species, in contrast to the more mundane and ubiquitous functions of previously identified operons that have to do with sugar metabolism or amino acid biosynthesis.

The exact nature of the links between the methylation of *tar* and *tsr* gene products and changes in flagellar rotation are unknown. However, a plausible scheme for the involvement of these genes in the later steps of the pathway has been suggested by Parkinson (1977). He notes that the closely linked *cheB* and *cheZ* genes—defined by mutants that tumble excessively—may code for cytoplasmic products that together interact with the MCP factors in the membrane. The *cheB* and *cheZ* factors would be important in carrying the information down the pathway, but would not directly control counterclockwise versus clockwise rotation of the flagella. This is because these two mutants can respond to increases in attractant, which will suppress their excessive tumbling. Four additional genes—*cheC, cheA, cheW,* and *cheY*—may control a subsequent step in the path, relating to a switch mechanism that produces the reversals in flagellar rotation. Mutations in these genes generally cause an inability to tumble, even in the presence of tumble-enhancing stimuli.

Products from the *cheA, cheW,* and *cheY* genes have been identified as cytoplasmic proteins of particular sizes; but the nature of the molecules' functions in the control of rotation and switching has not been determined. The rather direct involvement of the other locus,

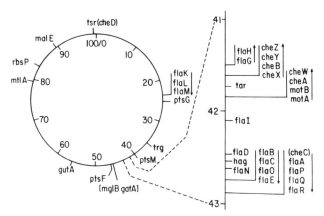

**Figure 17.**
Chemotaxis-related loci in *E. coli*. Approximate map positions (in minutes) of chemotaxis genes were taken from Bachman, Low, and Taylor (1976). Arrows above gene clusters indicate the direction and extent of cotranscription within the cluster. Genes within brackets have not been ordered with respect to outside markers. [Parkinson, 1977]

*cheC,* in flagellar functioning is dramatically revealed by the isolation of different mutant alleles of this gene. Most such mutants, in fact, have no flagella; but rarer *cheC* mutants (which are probably "non-null" mutants) have flagella that can only rotate counterclockwise (thus there is no tumbling). Certain "second-site" mutations that have been induced in a mutant *cheC* gene not only regain the ability to tumble, but they also tumble at an accentuated rate (Parkinson and Parker, 1979). Thus, the product of this gene could be a flagellar component involved in the switch mechanism.

Another mutation that apparently affects flagella is *cheU* in *Salmonella* (Khan et al., 1978; Rubik and Koshland, 1978). This remarkable variant is inverted in its chemotactic responses, being attracted to repellents and repelled by attractants. Another allele of this gene, *flaQ,* is nonflagellated, suggesting that the *cheU* mutation has an alteration in a specific component of flagella. The alteration in flagellar functioning is suggested to be a subtle change in response to intermediates in the pathway of chemotactic signal processing, leading to abnormally extended clockwise rotation (Khan et al., 1978). This actually converts the flagellar bundle into a right-handed helix due to changes in torsion. Smooth swimming is therefore associated with clockwise rotation of the flagella, and tumbling with counterclockwise rotation; this is the reverse of wild-type functioning and explains the inverted chemotactic behavior. Hence, there are profound behavioral consequences due to this simple genetic change; yet the behavioral abnormality can be understood without having to imagine that there are immense or bizarre changes in the sensory processing system.

The final evidence on the late steps in the pathway to be now discussed originates from genetic work on the interactions of the different products of the genes controlling these steps. Obviously, the number of different genes involved in chemotaxis does not necessarily specify the number of steps in the stimulus-transduction pathway because certain pairs of genes could interact to control a given step. The MCP-1 and MCP-2 factors coded for by the *tsr* and *tar* genes, respectively (see "Sensory Mechanisms"), may interact by being physically associated, since the dominant *tsr* mutants already mentioned cause the products of *both* of these genes to have little or no methylation. For the *cheB* and *cheZ* gene products, interaction is suggested by the fact that certain *cheB* alleles complement certain *cheZ* alleles poorly, though either mutant by itself is completely recessive (Parkinson, 1976). Such poor complementation could obtain if the products of these two genes form a protein complex.

There are apparently interactions between certain genes in putatively adjacent steps of the pathway. For instance, when one isolates revertants of the generally nonchemotactic *cheZ* mutants, the normally behaving strain that results often carries the original *cheZ* mutation and a compensating mutation in the *cheC* gene (Parkinson and Parker, 1979). Beginning with a *cheC* mutant, it is possible to generate a doubly mutant *cheC-cheZ* strain that behaves normally. These cases of complementing mutations are specific to the particular alleles involved and are not found if a mutation in either gene is a nonsense mutation (coding for no gene product) (Parkinson and Parker, 1979). The implications are that the presumably cytoplasmic product of *cheZ* and the flagellar component specified by *cheC* physically interact (see figure 3). Hence, an altered protein from one mutant cannot interact with the normal protein from the other gene; yet a mutation in the latter gene could lead to a specifically altered product that now will interact quasi-normally with the mutant protein from the former gene. The matter of second-site suppressor mutations that reveal interactions between genes of neurobiological interest also has appeared in the discussions of higher organisms.

## General Abnormalities of Motor Output in Higher Organisms

Two motor mutants from different eukaryotic species, tottering mutant in mouse and Hyperkinetic mutant in *Drosophila,* which are superficially similar to each other, serve to introduce important aspects of neurogenetic analysis of behavior in complex organisms.

*Tottering Mutant in Mouse*
The first mutant is "inherited epilepsy" in mouse, owing to the expression of an allele of the tottering (*tg*) gene (Green and Sidman, 1962). This is one of the host of mutants found in this mammal, initially on the criterion of general abnormalities (for example, Sidman, Green, and Appel, 1965). Young mutant mice expressing *tg* show abnormal bursts of spike waves as seen in electrocorticograms; the anomalous spikes are always found to be bilateral. Correlated with this abnormal activity are seizures involving sudden arrest of movement, plus twitching movements and jerks of the limb (Noebels and Sidman, 1979). The seizures occur one or more times per day "without warning," but they decline in frequency as the adult mice age.

The *tg* mutant was further studied to determine which parts of the brain may be involved in the attacks. This was done by injection of radioactive 2-deoxyglucose into mutant individuals during and between episodes of the spontaneous seizures, followed by autoradiography of brain sections. This well-known method for marking parts of the brain suspected of being involved in particular aspects of sensory input and processing, and in particular behavioral actions (Kennedy et at., 1975; Plum, Gjedde, and Samson, 1976; Sokoloff et al., 1977), has even been used to label areas of the brain thought to control circadian phenomena in the mouse (Fuchs and Moore, 1980; Schwartz, Davidsen, and Smith, 1980) and visual or auditory inputs into the brain of *Drosophila* (Buchner, Buchner, and Hengstenberg, 1979; Buchner and Buchner, 1980).

In the case of tottering, striking increases in uptake of 2-deoxyglucose uptake during seizures were observed, bilaterally, in particular portions of the brain stem (Noebels and Sidman, 1979). This labeling technique has therefore allowed a putative location of structures that are at least indirectly affected by the mutant gene. The 2-deoxyglucose experiment was important because no neuropathologies in labeled parts of the brain could be detected simply by sectioning brains from mutant individuals (Noebels and Sidman, 1979), though other alleles of *tg* have been reported to show morphological defects in the cerebellum (Yoon, 1969; Meier and MacPike, 1971; Tsuji and Meier, 1971; Nakane, 1976).

*Hyperkinetic Mutant in Drosophila*
The other motor mutant, Hyperkinetic (*Hk*) in *D. melanogaster* (Kaplan and Trout, 1969), has several motor abnormalities, which, as in the case of tottering, demonstrate some additional techniques for localizing tissues responsible for this kind of genetic defect. The *Hk* gene is one of the many in *Drosophila* defined by mutations leading to rather general motor defects (for example, Markow and Merriam,

1977; Homyk, Szidonya, and Suzuki, 1980). The abnormalities in Hyperkinetic mutants are among the best characterized.

The *Hk* phenotype, owing to either of two allelic mutations on the X-chromosome, involves leg shaking under ether anesthesia and also increased spontaneous activity (reviewed by Ikeda and Kaplan, 1974; Dagan, Kaplan, and Ikeda, 1975; Kaplan, 1979). The leg shaking is similar to, but distinguishable from, that caused by the Shaker (*Sh*) mutations discussed earlier (Trout and Kaplan, 1973). Data from intracellular recordings, including a direct comparison of the effects of *Hk* and *Sh* (M. Tanouye, unpublished), indicate that defects caused by these two mutations are fundamentally different.

The *Hk* mutations also lead to a spectrum of abnormalities that seem to be related to those just noted: in addition to the "primary" defects, mutant flies have an abbreviated lifespan and increased oxygen consumption (Trout and Kaplan, 1970), slightly aberrant mating behavior (Burnet, Connolly, and Mallinson, 1974), and a dramatic jump response to shadow stimuli (for example, Levine, 1974), which can even overcome the effects of a temperature-sensitive-paralytic mutation (Williamson, Kaplan, and Dagan, 1974). The jump response is often accompanied by a brief seizure of immobility in adults expressing only an *Hk* mutation.

Do all these behavioral changes in *Hk* have a single, unifying, underlying cause, or are they associated with simultaneous effects of the mutation in many tissues? In short, what is the *focus* of the mutant behavior, the anatomical site at which the mutant gene exerts its primary effect (Hotta and Benzer, 1972)? This question has been approached by using genetic mosaics (already introduced in the above discussion on neurochemical and physiological mutants). Because *Hk* was one of the first mutants analyzed in this way in *Drosophila,* and since the results were relatively straightforward, they will be briefly summarized in order to introduce the principles and techniques of this kind of mosaic analysis.

Since *Hk* adults have rhythmic bursts of impulses, induced by ether and recorded intracellularly from nerves in the thoracic ganglia (Ikeda and Kaplan, 1970a), one might suggest that the foci—one per leg—for shaking of these appendages are in the ventral nervous system. However, it could be that the brain contains the one focus for leg shaking, meaning that expression of *Hk* in abnormally active thoracic nerves is unnecessary. This hypothetical brain site could also be the one affected by the mutation to cause other abnormalities, such as the jump response to shadows. One way that this idea of unifying focus was ruled out was with mosaics involving somatic loss early in development of an X-chromosome from heterozygous *Hk*/ + zygotes (figure 18; see the earlier description of this technique in figure 15).

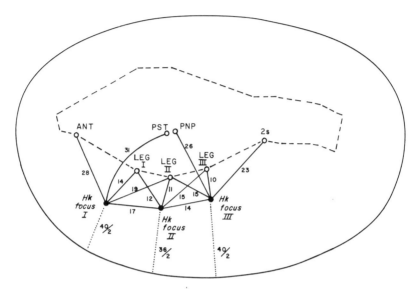

**Figure 18A.**
Fate-map sites of the behavioral foci for the Hyperkinetic mutant in *D. melanogaster*.
The dashed line represents an outline of the fate map of sites on the embryonic
blastoderm (Janning, 1978) that give rise to various parts of the adult cuticle (a few of
which are indicated: ANT, antenna; PST, presutural bristle; PNP, posterior
notopleural bristle; 2s, 2nd abdominal sternite; and LEG I, LEG 2, and LEG 3, legs on
the prothorax, mesothorax, and metathorax, respectively). The numbers on the map
designate the fate-map distances between Hyperkinetic foci and sites for cuticular structures and among the Hyperkinetic foci themselves. A given number represents a percen-
tage of the total number of mosaics analyzed in which, for example, the phenotype (and
genotype) of a cuticular structure (scored with respect to a body color marker mutation)
is different from the behavioral phenotype of a leg; hence, 14% of the mosaics involved
cases with LEG I marked but showing normal behavior or LEG I not marked but showing ether-induced leg-shaking behavior. Similarly, 17% of the mosaics were cases in
which the behavioral properties of LEG I and LEG II were different (I normal and II
shaking, or vice versa). The results indicate that there are separate focuses for each leg;
they fall in the region of the blastoderm that probably gives rise to the ventral thoracic
nervous system. [Hotta and Benzer, 1972]

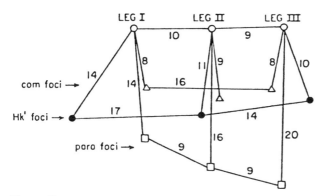

**Figure 18B.**
Foci for behavioral mutants of *D. melanogaster* with defects in leg behavior. The diagram is a fine structure map, referring to part of the thorax. Foci for ether-induced leg shaking, caused by the Hyperkinetic ($Hk^1$) mutant, are designated by ( ● ) and are the same foci shown in figure 18A. Each ○ is a fate-map site for the legs (as explained and pictured in figure 18A). The same basic strategy used for $HK^1$ was also employed in the mapping of foci for leg paralysis induced by the mutant comatose (*com*) at 38 °C ( Δ ) or by the paralytic temperature-sensitive mutation (*para*) at 29 °C (□). These foci were determined by Benzer (unpublished; cited in Hall, 1978a). [Hall, 1978a]

On the external tissues of each mosaic, mutant *Hk* tissues were identified by using a marker for the cuticle, while the *Hk*/ + tissues were heterozygous for the recessive cuticular mutation and were thus unmarked. These mosaics showed, first, that individual legs could shake independently in different mosaics (Ikeda and Kaplan, 1970b), instead of all six, or none, shaking in each separate mosaic. The genotype of the head was uncorrelated with the shaking of any leg; moreover, separate thoracic ganglia showed the abnormal burst of activity if the corresponding leg had shook, and other ganglia in the same mosaic fly had normal activity (Ikeda and Kaplan, 1970b). It has also been shown that a thoracic ganglion in an *Hk* fly, isolated by surgery from cephalic and sensory input, still has mutant activity. This result, however, does not prove that the foci are in the thoracic ganglia. These tissues could have been influenced by other, even nonthoracic, tissues during development.

Leg shaking caused by *Hk* and ether were studied further in mosaics by the fate mapping method (Hotta and Benzer, 1972), which involves placing sites on the embryonic blastoderm that develop into various parts of the fly. In these mosaics, the external genotype (regarding *Hk* of a leg was rather closely correlated with whether it shook; this is unlike what was found with respect to the genotype of structures of the dorsal thorax, head, or abdomen (Hotta and Benzer, 1972). More quantitative placement of the leg-shaking foci suggested that they correspond to progenitor cells for the ventral nervous system of the adult; that is, the foci mapped to a midventral part of the fate map shown in figure 18A (which explains the procedure used to locate the foci). Figure 18B also includes, for comparison, the foci for leg paralysis induced by physiological mutations discussed earlier.

Another behavioral focus associated with *Hk* has been mapped to a totally separate location from the leg foci; this is the site that is responsible for the jump response. W. D. Kaplan (cited in Hall, 1978a) mapped this focus to a part of the blastoderm near sites that develop into the cuticle of the adult head. *Hk* might, therefore, affect a ganglion in the head with respect to this particular feature of the mutant syndrome. Furthermore, mosaics showed that the control of *Hk*-induced jumps and seizures was, in a sense, unilateral (unlike what appears to be so for tottering in the mouse); this means that *Hk* mutant tissue in only one side of the head is sufficient to cause a mosaic to exhibit the full mutant behavior. This is a "domineering" focus, as opposed to a "submissive" one. The latter means that each of two bilaterally symmetrical sites must express the mutation in order that the phenotype of the whole mosaic be mutant. These terms are from Hotta and Benzer (1972), who also showed that the focus of a drop-dead mutation on the X-chromosome (which causes young adult flies to die) has a putative brain focus that is submissive, unlike the *Hk* focus mapping near or in this tissue.

Since visual stimuli are required to induce *Hk* mutants to jump, one might suggest that the optic system contains the focus. The eye would be excluded, since the genotype of the eyes in *Hk* mosaics was uncorrelated with mutant behavior. However, the optic lobes, just proximal to the eyes and flanking the brain proper, might contain the focus since both these lobes and the *Hk* focus map to a similar location on the fate map, separate from progenitors for the eyes (Kankel and Hall, 1976; W. D. Kaplan, cited in Hall, 1978a). Kaplan (1979), who has further examined the possible association of visual system defects with the jump response, noted that mutations (*rdgB* and *ora;* see "Sensory Mechanisms") that block input from outer photoreceptor cells of each ommatidia in the eyes to the first-order optic lobe also block the jump response when these visual mutants are combined with *Hk*. Yet a mutation (sevenless; see "Sensory Mechanisms") that blocks input from a central photoreceptor cell to the second-order optic lobe does not interfere with the jump response in the analogous double mutants. This might suggest that *Hk*'s focus for this behavior is in the first-order optic lobe (the lamina). But this is certainly not proved because the double-mutant experiments only show that visual pathways that are initiated in the outer photoreceptor and the lamina seems to be involved. The actual focus could easily be at a more central step of the pathway. Thus, the dissection of a mutational defect, by double mutants or by surgery, is not a substitute for mosaic analysis.

Other aspects of abnormal behavior caused by the *Hk* mutation have not been well studied in mosaics. Also, neither the foci in the thoracic ganglia nor in the brain have been studied with mosaic techniques that use an internal cell marker (cf. Kankel and Hall, 1976, and later discussion of mouse brain mutants analyzed in this way). But the mapping of two of the *Hk* phenotypes indicates that it is possible to gain clues as to where a putative neurological mutation affects the nervous system. It is not surprising that more than one neural tissue would be so affected, since different nerves throughout the animal should share many similar properties.

Other neurological mutations in *Drosophila* and in vertebrates have been powerfully analyzed with mosaics, though this approach is somewhat limited to the study of X-chromosomal mutations in *Drosophila* (see mosaic schemes in figures 15 and 30 and discussion in Hall, 1978a), and by the fact that only relatively large nerve cells can be marked in mouse mosaics (Mullen and Herrup, 1979). However, these limitations have still allowed several behavioral mutations to have their foci determined, and several of these genes have thus been shown, indeed, to affect the nervous system, a particular small part or cell type in it, or even several separate portions. This range of different foci has been found among several different kinds of behavioral mutants.

It will be valuable in future work to augment mosaic studies on a *Drosophila* mutant, such as Hyperkinetic, with 2-deoxyglucose "activity labeling" experiments (cf. Buchner and Buchner, 1980), just as it would be useful to have mosaic data on the sites of action of a mutation, such as tottering in the mouse. The mosaic approach, in either organism, has the advantage that it can lead to identification of primary foci, that is, cells and tissues directly affected by a given mutation. The activity labeling approach may identify such primarily affected sites and/or several other parts of the nervous system that are affected as well, not by direct action of the mutant gene in those cells, but, for instance, by interaction across synapses with the cells that are "at" the focus. A problem is that one cannot discriminate between primary and secondary defects by using the 2-deoxyglucose method, and this technique may not generate any useful data if it is used in conjunction with a neurological mutation that eliminates the presence of certain neurons.

## VISUAL BEHAVIOR

Genetic analysis of visual processing in *Diptera* has focused on the identification of individual components of a formal pathway and on the functional anatomy of the system in order to correlate physiological processes with anatomical pathways. These studies have been carried out in the context of a large body of behavioral, anatomical, and physiological information from wild-type flies of many species (for example, Götz, 1968, 1972; Reichardt, 1970; Dvorak, Bishop, and Eckert, 1975; Pierantoni, 1976; Strausfeld, 1976). The initial genetic approaches in this area have relied on the use of behavioral tests designed to probe selectively for different aspects of visual behavior, for example, sensitivity to dim lights and motion detection (reviewed by Heisenberg and Götz, 1975), followed by isolation and study of mutants in *Drosophila* that are defective in a subset of these tests. The second approach, localization of function, has made use of mutations that eliminate certain classes of cells and also of mosaic analysis of different kinds of neurological mutations. Together, these approaches have identified the retinal cell types feeding into various kinds of neural processing, suggesting an important role for lateral interactions of an inhibitory nature between different input channels in the second- and third-order optic lobes; they have also identified structures in the lobula plate and the posterior protocerebrum as essential steps in the pathway of response to motion in the visual field.

## Inputs from Peripheral Cells that Initiate Visually Triggered Responses in *Drosophila*

The compound eye of *Drosophila* contains three classes of photo-receptor cells, which are represented in each of the approximately 700 ommatidia of the eye and, therefore, cover the entire visual field. However, not all the cell classes participate in every aspect of visual processing. The three cell types are distinguished by their size, location in the ommatidium, sensitivity to different wavelengths of light, and the depth within the optic lobes to which their axons project (Pak and Pinto, 1976). The major cell type, known as $R_{1-6}$, consists of six cells that occupy the periphery of the ommatidium. Their rhabdomeres, the light-sensitive part of the membrane, are the widest and longest, extending the whole length of the ommatidium, and their axons terminate and synapse in the second-order optic lobe, the lamina (Braitenberg, 1972). The optical axes of $R_{1-6}$ in one ommatidium are not the same, that is, they do not receive stimuli from the same point in space, nor do they all project to the same site in the lamina. Rather, each of the $R_{1-6}$ cells in one ommatidium projects to different sites in the lamina, and each of these sites (cartridges), in turn, receives inputs from six or seven retinula cells, all of which have the same optical axis but which originate from separate neighboring ommatidia (Braitenberg, 1972; see figure 19). The other two cell types, $R_7$ and $R_8$, are represented by one cell each per ommatidium. Their rhabdomeres are

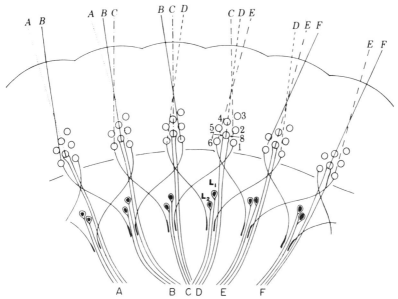

**Figure 19.**
Diagram of the connections of the retinula cells to the second-order neurons in the lamina of a Dipteran. Axons of peripheral retinula cells (1-6) having the same visual axis (*A, B, D, D, E,* or *F* indicated at the top of the figure) interweave to synapse on the large monopolar neurons (L$_2$, L$_2$) of a given laminar cartridge (A-F) indicated near the bottom of the figure. Axons of the central retinula cells (7 and 8) bypass the lamina and synapse in the medulla. For simplicity only the axons of cells 1, 6, 7, and 8 are illustrated (modified from Horridge and Meinertzhagen, 1970). [Pak and Pinto, 1976]

both narrower and shorter, occupying the central portion of each ommatidium. The rhabdomere of R$_7$ sits on top of R$_8$ longitudinally, and the microvilli of one cell are oriented orthogonally with respect to those of the other cell. The axons of R$_7$ and R$_8$ do not terminate or synapse in the lamina, but project through it, terminating in the next optic lobe, the medulla.

Differential functions had been postulated for these various cell types, based on their anatomical differences (Kirschfeld, 1972). The multiple representation in the lamina of each point in space mediated by the R$_{1-6}$ cells suggested that these channels should have a higher sensitivity than R$_{7,8}$. Another concomitant expected from this "superposition" of R$_{1-6}$ cells in the lamina was that any sensitivity to polarized light of individual retinula cells would be cancelled out, whereas R$_{7,8}$, with their orthogonally oriented microvilli, would potentially be capable of detecting polarization (discussed below). Furthermore, the narrowness of the rhabdomeres of R$_{7,8}$ should reduce the width of their receptive fields, allowing greater resolution and acuity of visual detection through these cells. Subsequent behavioral tests on houseflies seemed to bear out the postulate of two separate visual subsystems. The flies were tested for optomotor behavior, that is, the tendency to turn (measured as the torque of a stationary fly) in response to the motion in a horizontal direction of vertical black and white stripes in a visual field (for example, Götz, 1972). Various light intensities and widths of the black and white stripes were tested. Two kinds of response were obtained, one with high sensitivity but with poor resolution of narrow stripes (presumably mediated by $R_{1-6}$), and the other with high resolution, low sensitivity, and responsiveness to polarized light (presumably mediated by $R_{7,8}$; Kirschfeld, 1972). The hypothesis was further supported by the observation of different action spectra for the two kinds of response (Kirschfeld, 1972).

In *Drosophila,* a similar situation was expected to exist, given the high degree of anatomical similarity between the retina and lamina of the two species of fly (Harris, Stark, and Walker, 1976; Strausfeld, 1976; T. Hanson, unpublished). The hypothesis was bolstered when optomotor tests of wild-type flies revealed a similar association of high sensitivity with low acuity and, conversely, of low sensitivity and high acuity (Heisenberg, 1972). Furthermore, a number of mutants showed defects that seemed to be restricted to one "system" or the other (Heisenberg, 1972; Heisenberg and Götz, 1975). A definitive test of which cell types participate in the two kinds of visual perception was finally realized with mutations that selectively eliminate $R_{1-6}$, that is, outer rhabdomeres absent (*ora*) and retinal degeneration (*rdgB*), and a mutation that selectively eliminates $R_7$, *sev* (Harris, Stark, and Walker, 1976).

Heisenberg and Buchner (1977) studied the responses of these mutants and concluded that $R_{1-6}$ is the *only* cell type in the retina with input to optomotor behavior. The mutants *ora* and *rdgB* failed, for the most part, to respond to any light intensity or stripe width, whereas the mutant *sev* had virtually no change in its ability to resolve narrow stripes. The latter finding is especially significant in view of the fact that, in addition to eliminating half of the cells of the postulated subsystem, the absence of the rhabdomere of $R_7$ would greatly enlarge the receptive field of the remaining $R_8$ cell, impeding its ability to participate in high resolution (Heisenberg and Buchner, 1977). In further confirmation of this result, these investigators showed that optomotor behavior could also be abolished in flies with genetically normal retinula cells by selectively inactivating $R_{1-6}$ with bright blue light, a treatment that induces the prolonged depolarizing afterpotential (discussed in section on "Sensory Mechanisms"). Conditions that restore responsiveness to these cells (that is, orange light) also restore the optomotor response.

The cells $R_{1-6}$ have also been shown, by means of the same mutants, to mediate several other visually evoked behaviors (Heisenberg and Buchner, 1977; Wolf et al., 1980). These include the ability of a fly to fixate in flight on a vertically oriented stripe, the landing response to horizontal stripe motion, and the upward "thrust" response in flight to vertical motion of stripes (behavioral tests reviewed by Heisenberg and Götz, 1975).

Finally, sensitivity to polarized light in optomotor turning responses has routinely been assumed *not* to be mediated by initial reception through cells $R_{1-6}$, but, instead, by $R_7$ and/or $R_8$ (see arguments, based on the nature of the *E*-vector plane of polarized light, and the anatomy of the retinula cells in larger flies in, for example, Kirschfeld, 1967, 1971, 1973). These ideas were demolished, at least with respect to *Drosophila,* by the simple demonstration that polarization sensitivity is entirely normal in mutant sevenless individuals (Wolf et al., 1980). These authors discuss interesting possibilities for the adaptive significance of polarization sensitivity in visually evoked behaviors. They also note that, if the outer photoreceptor cells mediate such optomotor responses under differing degrees of light polarization, then sevenless mutants should behave normally; however, if $R_7$ or $R_8$ control these responses, the mutant should be insensitive or markedly modified, respectively, in its responses. It was not. Hence, Heisenberg and colleagues now seem, through their extensive and continuing use of visual mutants, to have eliminated the last vestiges of hypotheses suggesting that different kinds of optomotor responses are controlled, in part by the outer versus inner photoreceptors in the facets of these insect eyes.

It turns out that phototaxis is the only behavior to which $R_7$ and $R_8$ have been unequivocally shown to contribute (Harris, Stark, and Walker, 1976; Hu and Stark, 1980). It has been shown that flies lacking all retinula cells except $R_8$ (that is, the double mutant *sev rdgB*) retained phototactic behavior, though at a much reduced sensitivity relative to wild type. Furthermore, when $R_7$ is the only missing retinula cell, the phototactic response to ultraviolet light is greatly reduced, consistent with the spectral properties of $R_7$. These tests were performed under conditions known as fast phototaxis, in which the flies are agitated and then given 30 sec to choose between two different stimuli (one variable and one constant) in a Y maze (Harris, Stark, and Walker, 1976; Hu and Stark, 1980). Subsequent tests on the mutants *rdgB, ora,* and *sev* (Heisenberg and Buchner, 1977; Jacob et al., 1977) have extended these findings on phototactic input of $R_7$ and $R_8$ to slow phototaxis as well, in which flies have from 3 to 24 hr to distribute themselves between the stimuli.

## Interactions between Input Channels

Once information from the visual field has been received by cells in the retina, its analysis must necessarily involve interactions between elements of the visual system. In order to draw inferences as to the nature of such interactions and the levels at which they occur, genetic studies have been carried out to isolate different classes of interaction. From these studies have emerged direct indications of modulation of the $R_{1-6}$ channels by $R_{7,8}$ and some suggestions of interactions between separate elements of the $R_{1-6}$ system representing different parts of the visual field.

The first indication that input to $R_{7,8}$ could affect the functioning of the $R_{1-6}$ pathways came from optomotor studies on wild-type *Musca*. Kirschfeld (1972) focused beams of light on single retinula cells within an ommotidium and found that alternate stimulation of two cells of the $R_{1-6}$ type was sufficient to elicit an optomotor turning response in the restrained fly. However, this response could be abolished, or even reversed under certain conditions, if the rhabdomeres of $R_{7,8}$ were stimulated at the same time in a neighboring ommatidium (Kirschfeld and Lutz, 1974).

Genetic evidence from an interaction between the retinula cell types came from studies of slow phototaxis in the *Drosophila* mutants *sev* and *rdgB*, which permit selective stimulation of the different subsystems. Jacob and coworkers (1977) found that wild-type flies, which exhibit positive phototaxis under most stimulus conditions, will give a *negative* phototactic response to diffuse yellow light of sufficient intensity. However, the negative component of the response is completely eliminated in mutants lacking $R_{1-6}$ (*rdgB*) or $R_7$ (*sev*), implying that an interaction of the two systems is required to produce negative phototaxis.

Additional physiological and anatomical data support the idea that the interactions between $R_{7,8}$ and $R_{1-6}$ are inhibitory and suggest that they may take place in the medulla (see figure 20). Intracellular recordings from one class of lamina cells in the blowfly *Calliphora* have revealed that stimulation by UV light of a relatively wide area surrounding the receptive field of the cell will inhibit its response to stimulation in the center of the field (Mimura, 1976). As the primary UV receptor in the fly retina (Eckert, Bishop, and Dvorak, 1976; Harris, Stark, and Walker, 1976; Hardie, Franceschini, and McIntyre, 1979), $R_7$ is the most likely candidate as the mediator of this inhibition, at least at the level of initial light reception. The possibility that the anatomical basis of this phenomenon resides in the medulla is suggested by the fact that $R_7$ and $R_8$ do not make any synapses in the lamina (Campos-Ortega and Strausfeld, 1972; T. Hanson, unpub-

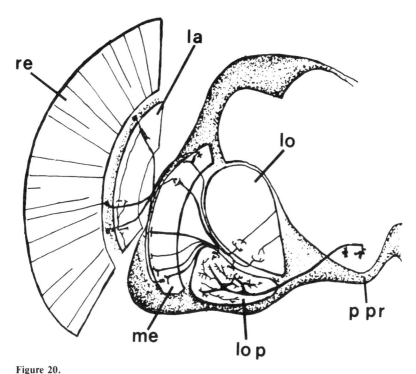

**Figure 20.**
Optic lobes and protocerebrum of an adult fly, illustrating anatomical structures relevant to optomotor behavior. The retina (re) and each of the optic lobes, the lamina (la),
medulla (me), lobula (lo), and lobula plate (lo p), are shown with schematic representations of some of the characteristic neurons and connections found in each lobe. The cortex containing the neuronal cell bodies (dotted area) surrounds the neuropil containing
the axons and synapses. Structures in which mutant tissue correlated with abnormal
behavior in the various mutants are the lamina, shown here with lateral connecting
neurons between the visual elements, and the lobula plate, shown here containing one of
the giant neurons that receive inputs through their extensive branches in that lobe and
that project into the posterior protocerebrum of the brain. The orientation of the
diagram is that of a horizontal section. [Greenspan, 1979]

lished) and that the inhibitory interaction can occur between two different ommatidia, eliminating any possibility of direct coupling between the $R_{7,8}$ and $R_{1-6}$ cells (if present at all) as a sufficient mechanism (see Schinz, 1978). Thus, a synapse seen in the medulla of *Musca* between $R_8$ and one of the large neurons originating from the lamina (Campos-Ortega and Strausfeld, 1972) may have some bearing on the interaction with $R_{1-6}$ (the lamina neuron receives direct input from the $R_{1-6}$ class of cells).

With the demonstration that optomotor behavior is mediated through $R_{1-6}$ in *Drosophila,* the narrow receptive field of $R_{7,8}$ relative to $R_{1-6}$ can no longer be used to account for the changes in resolution demonstrated by flies at different light intensities. However, there is reason to believe that the perception process is dissociable into a "high-acuity–low-sensitivity" component and a corollary "low-acuity–high-sensitivity" component, based on the existence of mutants that are selectively blocked in one of the two mechanisms. Thus, out of nine mutants that have either been isolated on the basis of poor optomotor behavior and phototactic behavior (Heisenberg, 1972; Heisenberg and Götz, 1975; Pak, 1975) or that were previously found to be deficient in visual function (Hotta and Benzer, 1970), four were defective in the ability to resolve broad stripes at low light intensity, while five were defective in the complementary process of resolving narrow stripes at high light intensity (Heisenberg, 1972; Heisenberg and Götz, 1975; Heisenberg and Buchner, 1977). Within either class, the mutants exhibit a range of response thresholds and maxima, but the two classes are quite distinct. Since they are both mediated by the same set of retinula cells, they presumably represent different adaptational states of the perceptual system. One mutant that is defective in the "high-acuity" system, *P-37,* exhibits a pattern of responses to various stripe widths at high light intensity that resembles the pattern of responses of wild-type flies to various stripe widths at low light intensity. Heisenberg and Buchner have suggested that this mutant, and others of its class, may be "frozen" in the dark adapted state.

The genetic identification of separate components of visual processing in these mutants is borne out by their differential responses in two other tests of visual behavior involving criteria independent of the optomotor test. The high-acuity-defect mutants are all severely impaired in their ability to fixate on and orient toward a vertically oriented stripe presented to the visual field, while the high-sensitivity-defect mutants respond relatively normally (Heisenberg and Buchner, 1977). In addition, the high-acuity-defect mutants have reduced sensitivity to UV light in phototaxis, while the other mutants have a somewhat increased UV sensitivity. The altered response to UV light may reflect disruptions in the components that mediate interactions between the $R_{1-6}$ cells and $R_{7,8}$ cells.

There is no information concerning where in the optic lobes these mutations exert their effects on optomotor behavior are located. However, it is interesting to note that the transients of the ERG (discussed above under "Neurotransmission") are abnormal in every one of the mutants (Hotta and Benzer, 1969; Heisenberg, 1971; Heisenberg and Götz, 1975) and that the first-order optic lobe, the lamina, is the most peripheral level in which lateral connections between visual elements are found (Strausfeld and Braitenberg, 1970).

Evidence in favor of a spatial separation, as well as, genetic separation, of the two components of stripe resolution derives from mosaic studies localizing the ERG defect in mutants. The ERG defect in the two mutants of high-acuity response that have been tested (tan and no-on-transient-A) was always associated with mutant tissue in the retina (Hotta and Benzer, 1970), while the ERG defects in the two mutants of high-sensitivity response that have been tested (ebony and extra-lamina-fiber, *elf*) did not correlate with the retinal genotype and may be associated with the lamina (Heisenberg and Buchner, 1977; Y. Hotta, cited in Hall, 1978a). The mutant *elf* also has an anatomical anomaly in the lamina: An extra monopolar neuron is found in each cartridge (Heisenberg and Buchner, 1977).

Independent genetic evidence concerning the involvement of the lamina in optomotor behavior has been obtained from studies of *Drosophila* mutants lacking acetylcholinesterase activity (*Ace* mutants). Greenspan and coworkers (1980) performed optomotor tests on flies that were mosaic for *Ace* mutations and found that, whenever mutant tissue was present in the lamina, the behavioral response was reduced or eliminated. These optomotor tests did not probe for subtle defects in sensitivity or acuity; rather, the parameters were those of wide stripe width and high light intensity, thus presenting the least stringent conditions possible. The lamina was not the only part of these mosaics that was mutant, but no cases of normal behavior were found in mosaics with *Ace*-minus clones in the lamina. Moreover, ERGs recorded from *Ace* mosaics revealed that the off-transient was eliminated in mutant laminas or mutant portions of mixed laminas (figure 17). If there is a connection between this physiological defect and the behavioral abnormalities, it may relate to the phenomenon of lateral inhibition observed in the responses of dipteran lamina cells (Arnett, 1972; Zettler and Jarvilehto, 1972; Mimura, 1976). Arnett (1972) recorded from lamina cells that, when stimulated at the periphery of their visual fields, gave no responses during the light stimulus, but produced discharges at light-off. If the off-transient of the ERG represents the summation of responses analogous to this, then *Ace* mosaics missing the off-transient might also be deficient in lateral inhibition, as if a class of cholinergic cells mediating collateral connections in the lamina (cf. Strausfeld, 1976, and figure 20) were selectively incapacitated.

## Central Aspects of Neural Functioning in the Optomotor Pathway

Despite the great success in using mutants to define roles of different peripheral cells in the control of various visual responses, it is still not known how visual stimuli, once received by *Drosophila* or the larger flies, are processed to result in responses such as optomotor behavior (cf. Reichardt, 1970; Götz, 1972; McCann, 1974; Buchner, 1976). However, additional genetic variants in fruit flies that disrupt particular portions of more central tissues may prove to be the principal tools for unraveling these more complicated features of the function of the visual system. What follows is a discussion of the initial phases of this analysis, in which studies of brain mutants have begun to augment those involving the eye mutants.

The importance of the lobula plate in visual processing was initially suggested by the physiological and anatomical characterstics of the neurons found in this part of the optic lobe in all species of *Diptera* examined. Cells responsive to motion in specific directions and to changes in the velocity of motion, as well as complex integrative cell types, have been found in this structure (for example, Bishop and Keehn, 1966; McCann and Dill, 1969; Mimura, 1971). Anatomical studies revealed giant neurons with wide-ranging dendritic fields, different classes of which ramify along particular axes of the lobula plate where they are presumed to contact the columns of cells topographically representing the visual field (for example, Braitenberg, Bishop, and Eckert, 1972; Dvorak, 1975; Pierantoni, 1976; Strausfeld, 1976). In combined studies, a class of giant neurons has been identified in the larger dipterans *Calliphora* and *Phaenicia,* which respond to horizontal movement in the visual field and whose dendrites arborize laterally across the lobula plate (H-cells); another class of giant neurons has also been identified, which respond to vertical movements and whose dendrites arborize in the dorsoventral direction (V-cells; Dvorak, Bishop, and Eckert, 1975; Eckert and Bishop, 1978). The possible relevance of these cells to behaviors, such as the optomotor turning response, is self-evident (Pierantoni, 1976).

The lobula plate of *Drosophila* also contains giant neurons homologous to those of the larger flies, that is, H-cells that arborize

laterally and V-cells that arborize dorsoventrally (Heisenberg, Wonneberger, and Wolf, 1978). Genetic evidence concerning the role of these
cells in behaviors that require perception of horizontal movement has
been contributed by studies of a mutant, optomotor-blind ($omb^{H31}$;
Heisenberg, Wonneberger, and Wolf, 1978), in which these cells are
absent or greatly reduced. The $omb^{H31}$ mutant is severely impaired in
its optomotor turning responses, exhibiting only about 10% of the
wild-type response. It is also defective in its ability to orient toward a
vertical stripe in the visual field (cf. Reichardt and Poggio, 1976). In
this case the defect is rather specific in that $omb^{H31}$ flies can fixate on a
stripe if they happen to be pointing toward it, but they fail to turn
toward the stripe if it is placed in a lateral position. They also
responded poorly to movements of the stripe. The interpretation of
these experiments with respect to the role of H- or V-cells is hampered
by the uncertainty whether $omb^{H31}$ affects other parts of the nervous
system aside from the giant cells of the lobula plate, an uncertainty
that might be resolved by analysis of mosaics. However, the relative
selectivity of behavioral defects induced by the mutation suggests that
it is not causing a generalized impairment of the visual system and also
indicates those behaviors for which H- and V-cells are probably *not*
required. The behaviors that are little affected in $omb^{H31}$ are the
upward-thrust response in flight to vertical movement of patterns
and the landing response to certain horizontally moving patterns
(Heisenberg, Wonneberger, and Wolf, 1978). These results indicate
that the mutant is not totally blind to movement. Moreover, the relatively normal thrust response indicates that the V-cells probably do
not mediate this behavior, and the performance of the landing
response indicates that horizontal motion detection is not totally
abolished.

Another optic lobe mutant with more extensive morphological
defects in these visual ganglia is small-optic-lobe ($sol^{KS58}$) (Fischbach
and Heisenberg, 1981). In spite of the severely depleted numbers of
neurons in the medulla and both lobular ganglia, and the miswiring
seen with respect to certain of the remaining cells, mutant individuals
are able to carry out a near-normal optomotor yaw response. However, the orientation behavior to a vertical stripe is impaired in $sol^{KS58}$
flies (Fischbach and Heisenberg, 1981).

One more mutation with relatively selective effects on visual behavior—though, as yet, with no known anatomical correlates—is the no-
fixation ($nof^{S100}$) variant of Heisenberg and Wolf (1979). These
mutant flies are quite abnormal in some features of behavior involving
fixation on objects or movements to and fro in a "choice" experiment between two identical objects (in which case $nof^{S100}$ individuals
are as defective as those that are totally blind); but in other behavioral

tests, such as those monitoring movement-induced course control, this mutant seems to be normal in its visual responses (Götz, 1980).

Additional support for the involvement of the lobula plate in optomotor behavior and new indications of additional requisite structures have come from mosaic studies of acetylcholinesterase (*Ace*) mosaics. A reduction or virtual absence of optomotor behavior in these mosaics is correlated with *Ace*-null clones in the lamina, the ventral lobula plate, and the posterior protocerebrum (Greenspan, Finn, and Hall, 1980). Although none of the mosaic patches in the animals tested was so exclusive as to consist *only* of one of these structures, there were no mosaics with mutant tissue in the optic lobes that performed normally, and the lamina and ventral lobula plate were the optic lobe structures that were mutant with the greatest frequency. Since these frequencies refer to sites in the cortex of the CNS where the cell bodies lie, it is interesting to note that in *Musca* the cell bodies of the H- and V-cells do lie in the ventral cortex of the lobula plate (Pierantoni, 1976).

The implication of the posterior protocerebrum in optomotor behavior extends the genetic analysis of the pathway into the brain proper. All of the optomotor-defective mosaics that had totally wild-type optic lobes had mutant tissue in mid-to-dorsal regions of the posterior brain. Many neurons of the lobula plate project to, and make synapses in, this region, and a number of neurons that descend from the brain into the thoracic ganglion originate here (Power, 1943; Pierantoni, 1976; Strausfeld, 1976; Heisenberg, Wonneberger, and Wolf, 1978). Some of these descending neurons are postsynaptic to the giant neurons of the lobula plate (Strausfeld, 1976; Strausfeld and Obermayer, 1976; see also figure 19). This part of the brain may represent the interface between inputs from the visual system and motor output in the thoracic ganglion.

The question how the two sides of the brain and visual system interact during these visually driven behaviors has not been addressed by studies of any of the mutants isolated on the basis of defective optomotor behavior or phototaxis. In optomotor tests on flies that are kept stationary with respect to the moving stimulus, stimulation of one eye will elicit an optomotor turning response (Gotz and Wenking, 1973), as will stimulation of the mutant sine oculis, which is missing one eye (Götz, 1970). When different stimuli are applied to each side, the fly still responds, though the direction and magnitude of the response depend in a complex manner on the stimulus parameters (Götz, 1972).

The results of optomotor tests on *Ace* mosaics, all of which had mutant tissue on one side only, were not comparable to any of the previous results; that is, *Ace*-null tissue in the posterior protocerebrum and/or the optic lobes of one side was sufficient to eliminate the response (Greenspan, Finn, and Hall, 1980). It should be noted that in the *Ace* study, optomotor behavior was measured in flies freely running under a watch glass, as opposed to the measurements of turning movements on stationary flies in the other studies, which may more easily reveal small optomotor responses. However, Greenspan and coworkers (1980) established that, even under their test conditions, flies stimulated through only one eye gave optomotor responses comparable to those of wild-type controls. These flies were mosaics with one eye made genetically blind by the *norpA*$^{EES}$ mutation (see "Sensory Mechanisms").

The results from *Ace* mosaics suggest that certain structures must be functional on both sides of the posterior protocerebrum *and* the optic lobes for normal optomotor behavior to occur. There are many inputs to the posterior protocerebrum and the lobula plate from the contralateral optic lobes and brain (for example, Mimura, 1971; Strausfeld, 1976) whose function may be required for coordinated motor output whether or not both eyes are receiving stimuli. The *Ace* defects may also be acting at a more peripheral level, such as the lamina. If the *Ace*-null laminas are defective in lateral inhibition, then the animals may be impaired in their detection of contrast or motion. Thus, the presence of an abnormal signal to the brain could override the signal from the normal side, and this defective input from one side may have more severe consequences than an absence of input from that side.

In conclusion, it seems as if the genetic dissection of central processing of visual information has involved rather *selective* effects of mutations on the anatomical, physiological, and behavioral features of this system. The specific defects seen in these preliminary studies and their subtle and surprising attributes have already been informative. It is not clear that these kinds of perturbations could have been achieved by surgical treatments or by injection of drugs.

## CIRCADIAN RHYTHMS

### Microbial Rhythm Mutants

Circadian rhythmicity underlies the control of many features of behavior and other biological phenomena, such as the eclosion of insects emerging as adults. Rhythmic oscillations occur not only in higher forms but also in microbial organisms, such as in the phototactic behavior in *Chlamydomonas* (reviewed by Mergenhagen, 1980). In fact, single gene mutations of this organism alter the circadian rhythms of phototaxis; these are the *per-1, per-2,* and *per-4* mutants, each caused by a change in a different gene and each having lengthened periods relative to the wild-type value, which is approximately 1 day (Bruce, 1972; Bruce and Bruce, 1978). There are mutants in other microbes, such as *Neurospora,* that affect rhythms (Feldman and Hoyle, 1973; Gardner and Feldman, 1980). In these cases a growth rhythm (period length $\approx$ 22 hr in the starting strain) was changed to a shorter or longer period length by mutations in at least two separate genes; four of the *frq* mutations turned out to be closely linked to each other and could be alleles of each other. This is interesting because some of the mutations in this "cluster" lead to abnormally short and others to abnormally long periods. This is not necessarily unexpected because of what has been found in the circadian clock mutants of *Drosophila.* There are potentially additional members of the *frq* gene cluster in *Neurospora;* these mutants, which are resistant to oligomycin, map very closely to the *frq-1, frq-2,* and *frq-3* mutations and, in addition, have shortened circadian periods (Dieckmann and Brody, 1980). These *oli$^r$* mutants may lead to an understanding of the molecular basis of some circadian rhythms because this gene in *Neurospora* appears to code for a small polypeptide in the membrane portion of the mitochondrial ATP synthetase (Sebald and Wachter, 1978).

Since the period length of circadian rhythms is altered in several of these microbial mutants, the suggestion has been made that some aspect of the basic timing mechanism of the circadian clock is changed. However, one of the *Neurospora* mutants, *prd* (formerly known as *frq-5*), which is unlinked to the other four mutations (Feldman and Atkinson, 1978), has a variety of biological defects. This pleiotropy leads to the idea that rhythmicity may be only indirectly affected by this mutation.

Mutations in other lower forms, for example, *Paramecium,* have been found that lead to the simple loss of rhythmicity (for example, Stadler, 1958; Barnett, 1966, 1969; Sargent, Briggs, and Woodward, 1966); but many "nonclock" functions can affect a behavioral or other biological phenomenon, and so the clock per se is not necessarily altered in an arrhythmic strain. (This important point will be addressed again in the discussion of circadian mutants in *Drosophila.*) Before considering such higher forms—in which the relation of rhythmic oscillations to neurobiological phenomena becomes important—it should be noted that oscillating biochemical systems have been discovered in lower forms such as yeast (Chance et al., 1973). Sometimes these rhythms, which have been demonstrated both in vivo and in vitro, oscillate with much shorter periods than do the circadian rhythms (for example, Chance et al., 1973). The relation of the former to the latter is still a mystery, and the biochemical rhythms in microbes do not appear to be of neurobiological interest. However, a striking example of a connection between mechanisms that control short-term and long-term fluctuations is considered from a genetic point of view in a later section on reproductive behavior in higher organisms. This discovery involves the analysis of rhythmic fluctuations related to auditory communication in the courtship of *Drosophila* (see section on "Reproductive Behavior").

## Rhythm Mutants in *Drosophila*

In insects there are genetic variants that disrupt circadian rhythmicity. These arthropods have regular oscillations occurring in a variety of parameters, such as those involved in egg laying, eclosion, general locomotor behavior, or reproductive behavior (reviewed by Saunders, 1976; Konopka, 1981). An appropriate question might be; Does the same clock control two or more oscillating phenomena in mutants in these organisms? Genetic selection experiments were one way this question was approached. Thus, in *Drosophila* and in the moth *Pectinophora,* different aspects of rhythmicity were found to respond differently to selection of alterations in one function. For instance, selection for lines with altered eclosion rhythms might not lead to differences in oviposition in the new strains (Pittendrigh, 1967; Pittendrigh and Minis, 1971).

Single-gene differences in *Drosophila* provide an even more useful approach for asking about the control of different oscillating phenomena. Konopka and Benzer (1971) have isolated three period (*per*) mutants in *D. melanogaster* based on criteria of altered eclosion rhythms; each mutant was subsequently found to affect circadian rhythms of locomotor activity as well (figure 21). Both kinds of rhythm are also affected in five arrhythmic mutants (defining two genes) that have been isolated in *D. pseudoobscura* (Pittendrigh, 1974; Konopka and Wells, 1980). It is also known that the mutations in *D. pseudoobscura* affect both phase response and periodicity per se, implying that both can be controlled by the same underlying factor specified by a given gene.

Of the three mutant alleles of the *per* locus in *D. melanogaster,* one (*per⁰*) is arrhythmic (Konopka and Benzer, 1971). Genetic and behavioral tests involving the construction of females completely deleted of the *per* locus (which readily survive) suggest that the *per⁰* allele "inactivates" this gene since individuals expressing this altered allele or deleted of the gene are similarly arrhythmic (Young and Judd, 1978; R. F. Smith and R. J. Konopka, unpublished). Hence, flies expressing *per⁰* may be totally lacking a molecular factor involved in a clock mechanism. But what if the absence of this factor indirectly and nonspecifically abolishes periodicity through some general defect or major lesion in the structure or functioning of part of the CNS? That

**Figure 21.**
Locomotor activity rhythms, monitored in infrared light, for individual rhythmically normal or mutant *D. melanogaster* previously exposed to LD 12:12. Activity registered by event recorder. Records read from left to right, each new line representing the start of a successive interval. For visual continuity, each successive interval is also replotted to the right of the immediately preceding interval. The traces for normal (*A*) and arrhythmic (*B*) are plotted modulo 24 hr; for the short-period mutant (*C*), modulo 19 hr is used; the long-period mutant (*D*) is plotted modulo 28 hr. Temperature is 25 °C. [Konopka and Benzer, 1971]

*per⁰* may be defective in the actual clock is strongly indicated by the fact that two other mutant alleles of this same gene, *perˢ* and *perˡ*, lead to abnormally short or abnormally long periods of rhythm (19 and 29 hr, respectively). Therefore, the mutations that do not simply eliminate oscillating phenomena are very useful in suggesting that the allele that does abolish circadian behavior is important. A biochemical search for the substance present in *per⁺* flies but absent from *per⁰* mutants may lead to the discovery of a molecular component in nerve cells that is directly involved in a clock mechanism.

Indeed, some data suggest that the major effects of the *per* mutants occur in the nervous system. Mutants of this X-chromosomal gene have been studied in the chromosome-loss mosaics described in more detail elsewhere in this book. The apparent focus of gene action (as defined in the previous section on motor mutants) was shown in the mosaic work to be close to the head of the insect (Konopka, 1972; Konopka, Wells, and Lee, 1981b). Transplantation experiments have also been performed, in which the brain from a *perˢ* adult was placed in the abdomen of a *per⁰* adult host. Periodic behavior corresponding to the phenotype of the rhythmic but mutant donor could be induced in the null host, in a fraction of the hosts surviving the operation (Handler and Konopka, 1979). Nonspecific corrective effects of the operation itself would appear to be ruled out by the fact that the characteristics of the specific mutant were transplanted into the host. A more trivial explanation might have been entertained if wild-type donor tissue had been used; thus, the importance of the availability of an array of mutant alleles of one gene, some without a null phenotype, is reemphasized. The transplantation results indicate that a specific diffusible factor produced in the nervous system is important in the action of the *per* gene.

Other information on the *per* mutant brains comes from a morphological study of neurosecretory cells in the posterior part of the fly's head; an abnormally large proportion of these cells has been found to be misplaced (that is, higher up in the brain than normal) in individiuals expressing a *per*° mutation (Konopka and Wells, 1980). This finding was also obtained in two aperiodic mutants of *D. pseudoobscura* (Konopka and Wells, 1980). Those neuroscretory cells may be part of the fly's circadian control of behavior.

It might be valuable eventually to perform neurobiological analysis of genetically altered circadian phenomena at the level of oscillations studied in single neurons (for example, in *Aplysia;* Strumwasser, 1965). The genetically tractable higher organisms, such as *Drosophila,* have not yet had "cellular circadian rhythms" studied in sufficient detail, though Konopka and colleagues have recently discovered a circadian rhythm of cell division in cultured wild-type cells of this insect (Konopka and Wells, 1981).

## LEARNING AND MEMORY

Information storage and retrieval by nervous systems are poorly understood, especially in terms of cellular and molecular mechanisms underlying conditioned behavior. Mutants with defects in learning and memory might provide a substantial increase in the strategies available for studying these higher functions. There are many genetic studies on learning in mammals, but most involve strain differences that sometimes result from selection experiments (for example, Bovet et al., 1969). These strains, however, are relatively intractable from a genetic standpoint.

### Behavioral Analysis of Learning and Memory Mutants

Learning and memory have been demonstrated in *Drosophila,* and this has led to the identification of some important individual genes that are involved in the control of conditioned behavior in this insect. Such learning behavior, mainly involving the fly's avoiding a stimulus that has been previously coupled with an adverse treatment, has been shown to be quantifiable. These treatments have involved the use of electric shocks (Quinn, Harris, and Benzer, 1974; Spatz, Emmans, and Reichert, 1974), noxious chemicals (Quinn, Harris, and Benzer, 1974), shaking (Bicker and Reichert, 1978), and unreceptive females (Siegel and Hall, 1979; see later section on reproductive behavior). Positive reinforcement (feeding) also works if the flies are hungry (B.

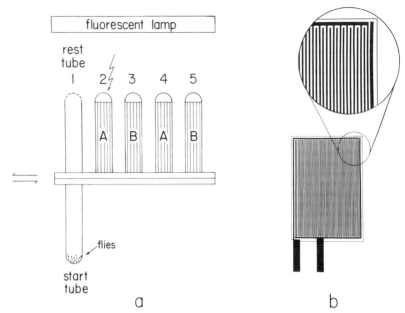

**Figure 22.**
Apparatus used for training and testing *Drosophila*. Two plastic blocks holding tubes slide on a dovetail joint so that the start tube can be shifted into register with tubes 1–5. Tube 1, the rest tube, is perforated for ventilation. Tubes 2–5 contain grids with odorants A or B (symbolizing two organic compounds, such as 3-octanol and 4-methyl-cyclohexanol). Tubes 2 and 3 are used for training. The lightening bolt indicates voltage on the grid. Tubes 4 and 5 are for testing, that is, for determining whether flies that had been shocked in the presence of odor A will then tend to avoid that odor, dispersed on the grid (right side of figure) in a testing tube but not associated with shock. [Dudai et al., 1976].

Tempel and W.G. Quinn, unpublished). The stimuli associated with the treatments have included specific odors (for example, Quinn, Harris, and Benzer, 1974), colors (for example, Quinn, Harris, and Benzer, 1974; Spatz, Emmans, and Reichert, 1974), and intensities of light (Bicker and Reichert, 1978).

In one example of a learning test, flies are shocked when they run into a tube in which an artificial odor has been placed (figure 22). Within a reasonable amount of time after such training, a significant fraction of these flies will tend to avoid a nonshock tube containing the odorant (purposefully used at a nonrepelling concentration) previously coupled with the adverse stimulus. This particular experimental strategy (Quinn, Harris, and Benzer, 1974) has been very important because all of the learning mutations have been isolated using this system.

The different stimuli that the flies will learn to avoid when they are coupled with something like shock are of some intrinsic interest. First, their effectiveness shows that the fly can discriminate between two different stimuli, such as different colors, in the first place. It is not trivial to demonstrate color vision, and learning has been an effective way to do this in *Drosophila* (for example, Menne and Spatz, 1977). Genetic experiments have augmented these demonstrations of color discrimination. For example, flies missing the "outer" photoreceptors from each ommatidium of their compound eyes, owing to the expression of a receptor-cell-degeneration mutation (see "Sensory Mechanisms") can still learn to discriminate different intensities of light, but they can no longer learn to tell different colors apart (Bicker and Reichert, 1978), probably due to the specific kind of visual input that is removed when these cells are selectively removed by the mutation (Harris, Stark, and Walker, 1976). A second reason for using an array of stimuli in learning experiments is that, when a putative mutant is found, one would like to find out whether it fails to learn on a variety of criteria or, on the other hand, the mutation merely disrupts a sensory modality such as sight or smell. Indeed, a mutant with defective olfaction was found on the basis of abnormal olfactory learning (Aceves-Piña and Quinn, 1979), but the other mutants found do not have defects in sensing external stimuli, or in performing a variety of simple and complex behaviors (for example, Dudai et al., 1976; Quinn, Sziber, and Booker, 1979).

These mutations that appear rather specifically to affect learning are in the X-chromosomal genes dunce, turnip, cabbage, and rutabaga. These mutants were isolated on the basis of near-zero learning in the shock odor system (Dudai et al., 1976; Aceves-Piña and Quinn, 1979; Quinn, Sziber, and Booker, 1979; Duerr and Quinn, 1981).

One use of these mutants has been to show that two seemingly different features of experience-dependent behavior can be controlled, at least in part, by the same underlying mechanism. Thus, whereas the mutants were found on the basis of defective conditioned behavior of adults, mutations also affect learning in larvae (Aceves-Piña and Quinn, 1979). Moreover, the dunce, turnip, and rutabaga mutations not only affect associative conditioning, a "higher" form of learning, but also habituation (Duerr and Quinn, 1981). This can be shown in wild-type flies by stimulating the tip of a foreleg with sugar, eliciting a mouthpart-extension response, and showing an eventual decrement in the response after repeated stimulation. For these three mutants, significantly fewer of the tested individuals show the usual decrease in response of the proboscis to the sugar stimulation, which, in wild-type individuals, persists for at least 30 min. These normal flies, in addition to exhibiting the relatively long-lived habituation to sugar, show shorter-lived sensitization to such a stimulus. That is, concentrated sucrose applied to the proboscis of *Drosophila* increases subsequent responses to tarsal stimulation for 2–5 min. In two of the learning mutants, dunce and rutabaga, the time course of sensitization is reduced, especially in dunce adults, in which sensitization lasts for less than 1 min. The turnip mutant individuals, however, were indistinguishable from wild type in these sensitization tests (Duerr and Quinn, 1981).

It was thought that some mutants showing no learning might, in fact, be learning initially, but forgetting so rapidly that no effects of experience can be shown. Dudai (1979) showed that dunce mutants can learn in the shock-odor system, in somewhat special circumstances involving testing of the mutants, by using the odor just previously coupled with shock, very soon after training. An additional way of suggesting that a nonlearning mutant might really be a "very short

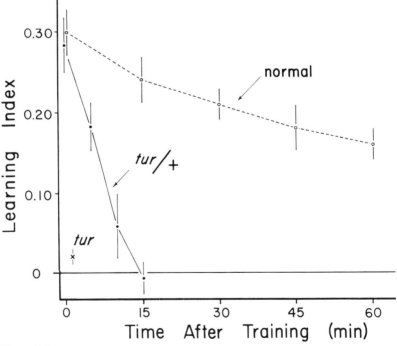

**Figure 23A.**
Memory retention in normal *Drosophila* (dashed line), turnip mutants ( × ), and turnip/ + heterozygotes (solid line). Populations of normal and amnesiac flies were given three training trials, left undisturbed for the periods indicated, then tested (procedure same as that in Quinn, Harris, and Benzer, 1974, except that flies were shifted to fresh tubes just before testing). The heterozygotes did not regain their memory later; learning indices at 20 to 60 min remained nero zero. [Quinn, Sziber, and Booker, 1979]

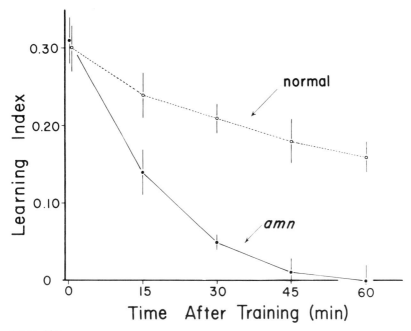

**Figure 23B.**
Memory retention of normal (dashed line) and amnesiac (solid line) *Drosophila*. See figure 23A for methods. [Quinn, Sziber, and Booker, 1979]

memory mutant" has involved testing of heterozygous turnip flies. This mutation is recessive, in that turnip/+ females learn normally. But the heterozygotes appear to forget what they have learned much more rapidly than do wild-type flies (figure 23A: Quinn, Sziber, and Booker, 1979), which, in fact, can retain some of their memory for as long as 6 hr (Dudai, 1977b). These "mild" effects of the mutation, seen when it is heterozygous with the normal allele, suggest that the severe effects in hemizygous males or homozygous mutant females are related to extremely short retention of the effects of training. There is another mutation that appears exclusively to affect memory. This is the X-chromosomal amnesiac variant (Quinn, Sziber, and Booker, 1979), which is recessive in its effects on learning or memory; but it causes hemizygous males or homozygous females to forget what they have learned relatively quickly (figure 23B), though not as rapidly as the decay seen in turnip heterozygotes. It should be noted that amnessiac individuals exhibit normal "retention" of the effects of sugar stimuli that result in sensitization or habituation, unlike what is found for certain of the other learning mutants (Duerr and Quinn, 1981).

With these learning and memory mutants in hand, one would like to discover how the genetic changes actually alter mechanisms of memory acquisition and storage. First, then, it must be shown with genetic mosaics that a mutant such as dunce or amnesiac has a defect of structure or function in the nervous system. It is not easy to perform such experiments because essentially all the learning tests in *Drosophila* have involved "population" phenomena, in which some fraction of the trained flies will avoid a stimulus previously coupled with aversive treatment, whereas a much lower proportion of the non-trained flies will avoid that stimulus (for example, Quinn, Harris, and Benzer, 1974). This feature of the *Drosophila* learning systems has inspired some vituperative but essentially unwarranted attacks on the utility of these genetic strategies (McGuire and Hirsch, 1977). Nevertheless, the population-learning tests are the ones that have been successfully used to isolate all of the mutants mentioned here. Several of these variants have been shown to be due to bona fide single gene mutations, which thus provide genetically uniform groups of flies for analysis, even at the biochemical level. Furthermore, the true learning mutants represent well-defined lesions that have been more useful in genetic studies of learning than the different "strains" of blowflies selected to have different learning abilities (McGuire and Hirsch, 1977).

Another way to compare mutant to wild-type flies has been simply to establish tests in which individual flies can reliably be shown to learn. Thus, Booker and Quinn (1981) have taught *Drosophila* to lift or to extend a leg in order to avoid electrical shock, as Horridge (1962) had shown in training experiments on individual cockroaches. Nearly all the wild-type flies learn correctly to perform this task but only if they are headless (cf. Horridge, 1962). By this method, mutant dunce, cabbage, and turnip flies were shown to be defective in the leg-lifting learning (Booker and Quinn, 1981). However, whereas a smaller proportion of the mutants, tested individually, learn to alter leg position to avoid being shocked, none of these three mutations results in every fly of that genotype failing to be conditioned. Indeed, up to 60% of the individuals learned to lift their legs in tests of one mutant (cabbage). One implication of this fact is that, if analysis of a series of individual genetic mosaics is attempted using the leg-lifting test, the only interpretable results will relate to the parts of the nervous system that are always mutant in mosaics that do *not* learn successfully.

*Biochemical Defect in a Learning Mutant*

One of the learning mutants may yield neurochemical information on the nature of defective conditioning that will go even deeper than the eventual determination of learning foci. This is dunce, the mutation that has been best characterized genetically, data on which have been pivotal in suggesting that the genetic analysis of learning may lead to an understanding of some of the molecular mechanisms involved in conditioned behavior. The dunce mutation has been mapped genetically to a very small subsegment of the X-chromosome (at the distal end, away from the centromere). Kiger and Golanty (1977) made a search of the *Drosophila* genome, using the technique of segmental aneuploidy discussed earlier, for genes affecting cyclic nucleotide phosphodiesterase activity. These enzymes and their control of cyclic nucleotide levels (cyclic adenosine monophosphate, cAMP, in particular) may be involved in experience-dependent phenomena in the fly, if one extrapolates from work on habituation and sensitization in molluscs (for example, Kandel, 1978). An X-chromosomal segment did have effects on cAMP phosphodiesterase levels (Kiger and Golanty, 1977), and this region is near the dunce locus. A homozygous deletion of the phosphodiesterase locus allows survival but causes female sterility (Kiger, 1977). Homozygous dunce females had been found to be subfertile or sterile, depending on mutant allele being tested. In fact, two dunce alleles were initially found by D. Mohler (Byers, Davis, and Kiger, 1981) on the basis of female sterility. Armed with this background information, Byers and coworkers (1981), Davis and Kiger (1981), and Kiger and colleagues (1981) have made the striking discovery that each of six independently isolated alleles of dunce, or the relevant homozygous deletion, leads to a severe reduction in one of the two forms of this enzyme activity that are present on sucrose gradients of wild-type *Drosophila* extracts (cf. Kiger and Golanty, 1979). The activity affected by this genetic locus is of a relatively low molecular weight and is rather specific in its activity against cAMP; the higher-molecular-weight species (unaffected by dunce variants) can also use cyclic guanosine monophosphate as a substrate. Flies expressing dunce mutations were shown not only to have reduced levels of a cAMP phosphodiesterase but also to have elevated levels of cAMP in homogenates of adults, as much as about sixfold more than the wild-type levels (Byers, Davis, and Kiger, 1981; Kiger et al., 1981).

Speculations on these biochemical results from dunce mutants, in light of the neurochemical work on cellular learning, are not yet warranted. However, since dunce affects habituation and sensitization as well as higher learning (Duerr and Quinn, 1981), mutant cells in the fruit fly may eventually be amenable to analysis in terms of specific physiological synaptic and molecular defects in cellular events that depend on previous experience.

## REPRODUCTIVE BEHAVIOR

### Courtship and Mating in Genetic Variants of Insects

Most of the genetic work on reproductive behavior has been done with *Drosophila* since courtship and mating in this genus have been analyzed from a hereditary point of view for a long time. Most of the earlier studies involved genetic selection and strain differences, interspecific comparisons, and mutations that alter the external morphology (for reviews, see Manning, 1965; Ewing and Manning, 1967; Grossfield, 1975; Ehrman, 1978).

*Mutants with Altered Courtship*
Single-gene mutations would also seem to be good tools for disrupting the fly in order to ask experimental questions on the control of reproductive behavior because the disruption might not be so general as to preclude meaningful performance of all aspects of these fixed action patterns. Some of the genetic variants to be discussed fulfill this purpose of making rather defined, limited changes in the CNS for assessing the effect of such "lesions" on courtship. Yet the variants are not so perturbed in their neural structure or function as to be devoid of all postures and movements. Since most of the relevant genetic variants appear to affect parts of the fly that are specifically involved in the control of reproduction, the discussion will not concentrate on mutants with defects in their external morphology or on complex strain differences.

Single-gene mutants affecting courtship have been isolated that seem to block different stages of the fruit fly's pathway of fixed action patterns (figure 24). Interstingly, there are no mutations that block the beginning of the sequence, that is, result in no courtship of a female by a mutant male. One might think that mutations with broader

**Figure 24.**
Pathway of fixed-action patterns in the courtship of *D. melanogaster*. Most of these behaviors are actions performed by the male, after he initiates courtship by recognizing the female and orienting toward her and then starts the pathway by tapping her abdomen with his forelegs. The courting pair may return to an earlier step of the pathway; for example, wing extension is frequently performed again after an unsuccessful attempt at copulation. The courtship mutants in this species (see the text) are placed above the sequence, at the stages that are blocked or otherwise disrupted by these genetic variants. [Hall et al., 1980b]

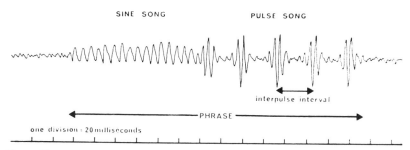

SINE SONG                    PULSE SONG

interpulse interval

PHRASE

one division = 20 milliseconds

**Figure 25.**
Waveforms in the courtship song of *D. melanogaster* males. The sine song component occurs in bouts of variable duration. The vibration component consists of trains of pulses, each pulse probably corresponding to one up-and-down movement of the wing. A phrase consists of one or more bouts of sine and pulse song. [Burnet, Eastwood, and Connolly, 1977]

effects than changes in courtship, such as those that are blind or cannot smell, would be behaviorally sterile; but this has not been shown. Further along in the courtship sequence, there is the cacophony (*cac*) mutant of Schilcher (1967a, 1977), which has an abnormal courtship song. These song patterns are produced by a species-specific series of wing vibrations, directed by the male at a female. Instead of showing the usual interval between pulses of tone of approximately 35 msec (figure 25), *cac* males exhibit more than 40 msec between the midpoints of individual pulses. The pulses themselves are also abnormal (for instance, louder), but other aspects of song, such as the hum produced before the trains of pulses (figure 25), are normal, as is wingbeat frequency. The mutant was found on the basis of poor mating; yet the abnormal song does not appear to be the sole cause of the generally poor mating, because wingless *cac* males still perform less well than wingless wild-type males (Schilcher, 1977). It should be noted that males without wings are still able to mate, although it takes them longer to achieve copulation (Ewing, 1964). Later, several other examples of a component of courtship behavior will be encountered in which a genetic treatment is found to cause a significant decrement in courtship but does not eliminate it. Hence, as might be expected, several components of the fly's nervous system and its behavior will be individually shown to contribute to mating success.

There is one additional component in the courtship song of *D. melanogaster,* not recognized in the early studies, that may make important contributions to sound communication. This involves the finding of a fluctuating interval between individual pulses of tone; the interval is not a constant 35 msec but shows a sinusoidal pattern of oscillation. The peak-to-trough amplitude is about 4 msec, and a full cycle of this variation for normal males lasts about 60 sec (Kyriacou and Hall, 1980). This discovery has led to the demonstration of a possibly unexpected connection between the genetic control of circadian rhythmicity and behaviors that oscillate with short period lengths. Kyriacou and Hall (1980) tested the three period (*per*) mutants of Konopka and Benzer (1971) and found the following: Strikingly, in males expressing the short-period allele (which leads to 19-hr circadian rhythms) the cycle of interpulse-interval variation is only 40–45 sec; the long-period allele results in lengthened cycles of 80–90 sec (cf. its 29-hr circadian rhythms); and *per$^o$* males have statistically arbitrary patterns of changing intervals that are neither sinusoidal nor are related to any other kind of function (figure 26). Therefore, the *per* gene seems to control a "major" clock in the fly's nervous system, one that controls rhythmical variation in behavioral phenomena occurring over hours and days as well as over minutes and seconds.

The period lengths of song rhythms of wild-type males have been found to be temperature-compensated (that is, they stay 1 min over approximately a 20 °C range; Kyriacou and Hall, 1980), in the same way that diurnal rhythms are a constant 24 hr when locomotor activity of *D. melanogaster* adults is monitored at a variety of temperatures (Konopka, Pittendrigh, and Orr, 1981a). However, it is interesting that the *per$^s$* and *per$^l$* mutants exhibit altered diurnal rhythms of activity at different temperatures, in that *each* mutant is more normal when monitored in relatively cold conditions (Konopka, Pittendrigh, and Orr, 1981a). It may be possible to show that the shortened and lengthened rhythms, respectively, of interpulse interval fluctuations in these two mutants show the same kind of temperature-induced instability when tested for oscillations of courtship song.

Courtship song oscillations are under interspecific genetic control, in that *D. stimulans* males, which are closely related to *D. melanogaster* males, have approximately 30-sec period lengths but larger amplitudes of variation (about 7 msec). The *cac* mutant of *D. melanogaster,* with its interpulse intervals similar in length to those of *D. stimulans* males, has a normally fluctuating song (for example, 60-sec periods); this reemphasizes the highly selective song defect caused by this mutation.

It is not known whether fluctuating intervals between song pulses are of adaptive value, for example, playing a role in species recognition. However, tests of the effects of simulated song (cf. Schilcher, 1976b) without oscillations or of *per$^o$* males with their arbitrarily changing interpulse intervals may reveal quantitative defects in mating success. The effects are not likely to be all-or-none because, as noted earlier, wingless males, given enough time, are able to mate.

Some interspecific work on courtship songs has shown that

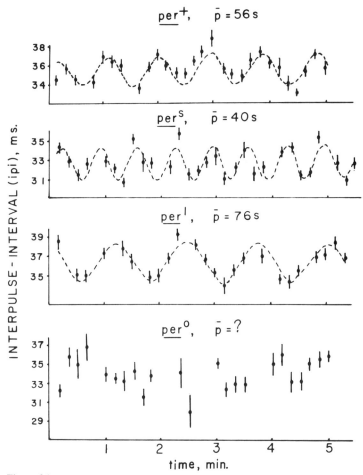

**Figure 26.**
Variation in interpulse interval for courtship songs of *Drosophila* males. *D. melanogaster* males that were wildtype (*per*⁺) or expressed a mutant allele of the *per* gene (see the text) had the wing vibrations they directed at females recorded. Mean intervals between pulses of tone (ipis) were calculated for successive 10-sec fractions of the 5–6-min observation periods. Each point represents such a mean ipi (in milliseconds = ms) together with the standard error. Points do not appear for all time frames monitored because males often broke off their courtship during some of the 10-sec intervals. A nonlinear regression was fitted to the points by the method of least squares, and the best fits to the data are shown as dashed lines. For *per*⁺, *per*ˢ, and *per*ⁱ males, sinusoidal functions gave highly significant reductions in the total variation ($p < 0.01$ in each case). However, the data from the *per*⁰ males gave a very poor fit to both sinusoidal and other functions. The period ($p$) for each of the sine waves was computed from the regression equation. [After Kyriacou and Hall, 1980]

"hybrid" songs can arise from males produced by a cross between two closely related crickets (for example, Hoy, 1974). Interestingly, the $F_1$ hybrid female, the cross of these same two species, responds best to the song from a hybrid male, as opposed to that of the male from either parental species (for example, Hoy, Hahn, and Paul, 1976). Thus, the control of the neural circuitry involved in the production and reception of the song may be under the control of similar components of the genotype of these insects. These sorts of tests on a well-defined genetic difference, such as *cac* versus *cac⁺* in *Drosophila,* may show that females carrying the song mutation respond best to the altered song from *cac* males.

At later stages of the courtship pathway, the celibate (*cel*) and fruitless (*fru*) mutations block attempted copulation (figure 24); thus, these mutants are behaviorally sterile (unlike all other mutants to be discussed here). The *cel* mutation was isolated on the basis of sterility and later found not to attempt copulation because of a single-gene defect on the X-chromosome (Hall et al., 1980b). The fruitless mutant (Gill, 1963), in addition to its defective interactions with females, induces unusual interactions among males.

At the latest stages of reproductive behavior in *Drosophila,* copulation time is rigidly programmed because its duration is invariant within a given species (Fowler, 1973). The male may be the sex that is genetically determined to exert the main control of copulation time because selection for strains with shorter or longer then normal duration (that is, 20 min in *D. melanogaster*) revealed that the males in the two divergent lines were responsible for the altered behavior (Mac-Bean and Parsons, 1967). A single-gene mutant that also alters copulation time in this species, coitus interruptus (*coi*), copulates for about 60% of the normal time. But *coi* also leads to abnormalities of sperm production or release (Hall et al., 1980b); thus, it may affect copulation time only indirectly. Another mutant affecting the final step in the pathway is stuck (*sk;* Beckman, 1970). Mutant males frequently cannot disengage from wild-type or *sk* females, or at least they take much longer than normal to pull away. When they do, various parts of their external genital aparatus are in abnormal positions. Yet there is no apparent abnormality in the morphology per se of the external structures before or after a copulation involving the mutant male (Hall et al., 1980b). Therefore, *sk* may be a neural or muscular mutant instead of one that is defective simply in the structural or mechanical aspects of the fly's genitalia.

*Mosaics Tested in Courtship*
Sex mosaics or gynandromorphs have also been valuable in breaking down both the behavioral steps and also the neurobiology of repro-

duction. (These gynandromorphs will be discussed in the section on "Developmental Neurogenetics" with respect to embryonic "fate mapping" though those studies are not concerned with the effects of sexual dimorphism in these mosaics.) The basic features of sexual behavior in gynandromorphs of *Drosophila* and other insects have been reviewed (Manning, 1967a; Hall, 1978a). In-depth observations of such mosaics were first done by Clark and Egen (1975), who tested many gynandromorphs of the wasp *Habrobracon* and found that their sex-specific behavior was controlled by the genotype of the head, presumably by the neural tissues in that body segment. Both male-specific and female-specific behaviors could be examined. For instance, mosaics with male tissue anterior in the head courted females but did not attempt the female-specific behavior of stinging prey; mosaics with female tissue in the head did not court in a malelike fashion but did attempt to sting caterpillars. A minority of these wasp mosaics showed the anomalous but interesting behaviors of both attempting to court females and also to sting prey. Perhaps they had mixed male-female brains (for example, one side male, the opposite side female), implying that only a portion of the brain needs to be of the relevant genotype for a particular sex-specific behavior and also that there is, at least, a partially separate and autonomous control of behavioral responses appropriate for a given sex. Other cases of unexpected behavior in Clark and Egen's mosaics were some rare individuals described as "wires crossed" because such wasps exhibited male courtship at a caterpillar or attempted to sting a female. Thus, it seems that the brains of some mosaics in insects may be abnormal, owing to a unique kind of neuronal miswiring that may occur at interfaces between male and female cells within a mixed ganglion.

Hotta and Benzer (1976) followed the work on wasps with a fate-mapping analysis of sex behavior in *Drosophila* gynandromorphs, which, like the wasp mosaics, were marked for male and female tissues only on the cuticle. In the fruit fly, the control of active male-specific actions directed at females was correlated with the genotype of the head, as is also the case in wasps. However, Hotta and Benzer (1976) found that only the earlier stages of the courtship sequence were controlled by anterior male genotype; this is unlike the situation in *Habrobracon,* where a male head is apparently sufficient for all the steps in male courtship to be carried out (except of course that a mosaic cannot copulate with a female unless the former has male genitalia). Positive female actions in the *Drosophila* gynandromorphs were not analyzable by Hotta and Benzer (1976) since, in *Drosophila melanogaster* at least, a female needs to perform no actions at all in order to be courted.

More recent work on courtship in *Drosophila* mosaics has involved

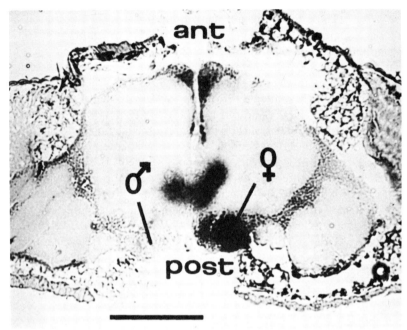

**Figure 27.**
Mosaicism in the brain of a male-behaving *Drosophila* gynandromorph. This is relatively dorsal, horizontal section (10 μm thick) through the brain of a mosaic that tapped, followed, and extended both wings at a female. (There was neither licking behavior nor attempted copulation.) The unstained cell bodies in the brain cortex are male, the stained tissues female. The anterior parts of the brain (ant) were bilaterally female in this and other dorsal sections. The posterior sites (post) were female on the right side, male on the left. The scale bar represents 100 μm. [Hall, 1979]

scoring the genotype of various ganglia in the gynandromorphs after analyzing them for male-specific or female-specific behavior. These mosaics, produced by somatic loss of an X-chromosome early in development, have had the haplo-X male tissues marked with respect to the histochemical expression of an enzyme mutation that does not perturb behavior (Hall, 1977, 1979). Behavioral and then histological analysis of these mosaics have shown that, if a sex mosaic exhibits any male behavior, the dorsal, posterior brain is always male on one side or both (Hall, 1977, 1979) (figure 27). But the courtship sequence is incomplete if only this portion of the brain is male. Male tissue in the thoracic ganglion is necessary for normal courtship song (figure 28), though wing display is controlled by the brain (Schilcher and Hall, 1979). Attempted copulation also requires male cells in the thoracic ganglia (Hall, 1977, 1979; cf. Hotta and Benzer, 1976). Male proboscis extension, which occurs after wing display and song but before attempted copulation, is controlled by the brain, as are earlier steps in the pathway (including tapping of females and orientation toward

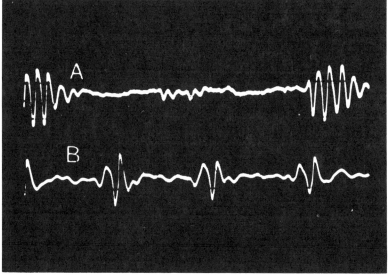

**Figure 28.**
Courtship song in *Drosophila* gynandromorphs. (*A*) Abnormal song from a mosaic with genetically mixed thoracic ganglia; there are no particular interpulse invervals, nor are there regular pulses of tone. (*B*) A normal pulse song, with approximately 34-msec interpulse intervals. [Schilcher and Hall, 1979]

them; see figure 29). However, male tissue is almost always required in both sides of the brain (that is, left and homologous right portions) if licking is to occur (Hall, 1979; cf. Cook, 1978).

A substantial fraction of the *Drosophila* mosaics that do court as males are quantitatively defective in the vigor of their performance, irrespective of whether they are blocked at a particular stage of the sequence. Moreover, individual gynandromorphs show highly correlated low courtship values in repeated tests (Hall, 1979), implying that the various brains that are part male, part female are characteristically defective as well as simply mixed in sexual genotype.

Preliminary analysis of female receptivity in gynandromorphs suggests that anterior tissues, possibly neural ones, must be diplo-X if a mosaic that is courted *and* has female genitalia will be receptive to copulation (Hotta and Benzer, 1976). It has been confirmed through the use of the internal cell marker that the brain apparently has to be female if a mosaic is to be receptive (L. Tompkins, unpublished). But a significant fraction of such mosaics that also possess normal female genitalia do not copulate; thus, another tissue may also have to be female for normal receptivity, such as an endocrine structure in the thorax.

What do these sex-specific neural foci mean? It could be that the intracellular properties of male and female neurons in these key

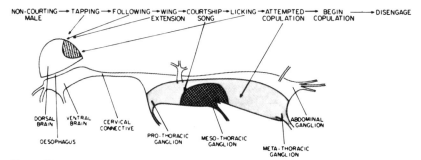

**Figure 29.**
The courtship pathway in *D. melanogaster,* analyzed in gynandromorphs. The parts of
the pathway controlled by different parts of the nervous system are shown by arrows
from the behavioral step to a part of the CNS shown here in sagittal section. This
breakdown of the pathway is from the analysis of courtship mosaics studied by Hotta
and Benzer (1976), Hall (1977, 1979), Cook (1978), and Schilcher and Hall (1979). For
the foci in the head that are shown in approximately the same portion of the brain, the
control of tapping, following, and wing extension (diagonal lines) are all domineering,
while the control of licking (stippling) is essentially submissive (see the text). The
thoracic focus for courtship song (cross hatching) is domineering. The focus for
attempted copulation (shading) is diffuse in that no particular portion of the thoracic
ganglia was male in every mosaic performing this behavior (though there was always
male tissue somewhere in the ventral nervous system in this category of gynandromorph). [Hall, 1979]

portions of the CNS are different; for example, they could possess different endogenous patterns of bursting activity. These cellular differences could be partially responsible for the fact that males and
females respond differently to stimuli from another individual.
Another possibility is that the connectivity among neurons may be
sexually dimorphic in the region of the neural foci detected by mosaic
analysis. Sex-specific differences of certain synaptic connections are,
in fact, observed in parts of vertebrate brains that are involved in
reproduction. Recently, sexual dimorphism at the anatomical level has
also been reported in male and female flies, in the optic lobes of larger
diptera (Hausen and Strausfeld, 1980; Strausfeld, 1980). The location
of this dimorphism may be near to or, at least, partially overlapping
with the courtship focus for the initiation of male behavior in
*Drosophila* (Hall, 1979). However, anatomical sexual dimorphisms
have not yet been detected in these smaller flies.

Yet another line of thought on the meaning of male-specific and
female-specific structure and function of the brain comes from work
on acetylcholinesterase mosaics in *Drosophila*. That gynandromorphs
in the fly exhibit male courtship whenever a part of the dorsal brain is
haplo-X may imply that one side of this portion of the brain need only
be nonfemale, instead of having to express any hypothetical male-specific functioning of the CNS. This idea stems from the fact that the

AChE gynandromorphs, which have no AChE in male tissues, still can show some courtship when mutant clones are in the dorsal, posterior brain (Greenspan, Finn, and Hall, 1980). These haplo-X tissues are neurochemically and morphologically defective, suggesting that non-female tissues, whether normally male or not, can, in effect, release male courtship behavior. This would be loosely analogous to the release of copulatory behavior in male praying mantises, which can be effected by ablation of a particular portion of the brain (for example, Roeder 1967). The hypothesis can be tested genetically in *Drosophila*, by constructing *Ace* mosaics that are thoroughly female (that is, not gynandromorphs); some of these mosaics will have the defective *Ace*-null tissue in the dorsal brain, and such females may have their latent ability to court other females released.

Parenthetically, it should be noted that the required genetic variants exist for the construction of choline acetyltransferase gynandromorphs, which would be *Cha⁻* in male tissues (Hall, Greenspan, and Kankel, 1979; Greenspan, 1980; see figure 15). This will provide a useful behavioral tool for the study of *Cha* mosaicism in the brain; no reliable histochemical stain is available for this enzyme, but courtship by a *Cha⁻* gynandromorph would positively identify this fly as possessing *Cha⁻* tissue in at least the dorsal posterior brain. Possible morphological defects, following development of a CNS that had lacked this critical enzyme, can therefore be assessed in the absence of a technique to stain directly the relevant mutant tissue.

If there is male-specific functioning of the brain in courtship, instead of merely a release of female inhibition as just suggested, then the location of the male behavioral focus in *Drosophila* is of interest. This is because it is near or possibly coincident with the origin of axons that innervate part of the "mushroom bodies" in this insect. These particular axonal bundles, and the various lobes they make up, apparently process a great deal of the olfactory input into the insect brain (Howse, 1975), and the mushroom bodies have been implicated in the control of specific courtship actions (reviewed by Huber, 1965, 1967, 1974; Elsner, 1973; see also recent evidence from Wadepuhl and Huber, 1979). Thus, in the fruit fly, the wiring or the intracellular activity of these groups of axons may have male-specific features related to the processing of sex pheromones. However, it must be noted that a mutant with defects in the *Drosophila* mushroom bodies, that is, mushroom-bodies-deranged (*mbd*), has been found to court normally in tests of males expressing this X-chromosome mutation (Heisenberg, 1980). Yet *mbd* does not result in the removal of all fibers in the mushroom bodies; thus, a definitive test of the effects of this gene on courtship must await the isolation of an allele with more severe anatomical defects.

*Chemosensory and Visual Control of Reproductive Behavior*
Whether the mushroom bodies are involved in controlling courtship, olfactory stimuli have been implicated in these behaviors for some time (Sturtevant, 1915; Shorey and Bartell, 1970; Averhoff and Richardson, 1974). Some recent work in this area has made important use of genetic variants. First, it has been shown that a gynandromorph will stimulate courtship as performed by a normal male if (and usually only if) the abdomen is at least partially female (Nissani, 1975, 1977; Hotta and Benzer, 1976; Hall, 1977; Cook, 1978). Jallon and Hotta (1979) generated a focus, by fate mapping, for such "female sex appeal" and found it to be relatively near the abdomen, but not at the abdominal surface. An internal structure, or possibly diffuse tissue in the abdomen, would seem to require the female genotype for sex appeal to be normal. However, Jallon and Hotta could not rule out the involvement of the posterior thoracic nervous system in the control of this aspect of courtship because they had used only external marking of their mosaics. L. Tompkins (unpublished), though, has been able to eliminate the possibility of CNS control by showing that a sex-appeal-positive gynandromorph can have an entirely male nervous system internally.

The abdomen is a common source of pheromones in insects (for example, Barth and Lester, 1973). Such materials from *Drosophila,* with their presumed abdominal source, have been detected by gas chromatographic analysis and by bioassays of the substances as they affect wild-type or mutant males in courtship behavior. Males and females have different volatile compounds in the chromatograms (Hedin et al., 1972). The materials from females, but not from mature males, stimulate courtship between two wild-type males if these pheromones are placed in a chamber with the males (Tompkins, Hall, and Hall, 1980; Venard and Jallon, 1980) or are as much as 0.5–1 cm removed from them (Tompkins, Hall, and Hall, 1980). The compounds from females stimulate courtship between *D. stimulans* males, which are closely related to, and will mate with *D. melanogaster;* yet the more distantly related *D. hydei* males are unaffected by the *D. melanogaster* female pheromones (Tompkins, Hall, and Hall, 1980). Further information on the importance of these sex pheromones is provided by tests of mutant males defective in olfaction. The smell-blind (*sbl*) mutation of Aceves-Piña and Quinn (1979) has been shown to eliminate the response of the flies in the bioassay tests of the volatile compounds from females; such mutant males do not respond to these compounds at any concentration (Tompkins, Hall, and Hall, 1980). Moreover, *sbl* males court less well than wild type, frequently appearing to orient improperly toward females and showing a twofold drop in quantitatively measured courtship activity. Venard (1980) has shown that another olfactory mutant (*olfC* of Rodrigues, 1980) is defective in its male courtship behavior.

An interesting parallel between these findings on olfactory-deficient flies and other results with chemotaxis-deficient nematodes is worth mentioning. Lewis and Hodgkin (1977) found that several mutant roundworms with reduced male potency were nonchemotactic. Similarly, Albert and coworkers (1981) found that three chemoreceptor mutants isolated by using a screen having nothing to do with male potency nevertheless showed reduced, but nonzero, mating efficiency. In one of these mutants, *che-3,* the male was found to be poor at sustaining contact with the hermaphrodite. Thus, it is possible the chemoreception plays a role in nematode mating as well as in *Drosophila.*

Olfactory-deficient flies are not behaviorally sterile; they simply court less well (cf. previous discussion of wingless males). Thus, it would appear that visual stimuli from female to male may be *sufficient,* at least in laboratory tests, to trigger the initiation of courtship and allow it to proceed with a reasonably high chance of completion. In *D. melanogaster,* though, males and females will mate in total darkness, eliminating the idea that vision is necessary in courtship. However, the role of this sensory modality has been reexamined more carefully through the use of visual mutants. It has been shown that blind males, such as those expressing *norpA* mutations discussed earlier and the glass-eye mutation discussed later in a developmental context, will mate with females (Siegel and Hall, 1979). But the behavior of these males is qualitatively and quantitatively defective, in a manner similar to the abnormal behavior shown by olfactory-deficient males (Markow and Manning, 1980; Tompkins, Hall, and Hall, 1980; Tompkins et al., 1981). A more subtle aspect of visually defective courtship is revealed in tests of the optomotor-blind mutant (*omb^{H31}*), discussed previously with respect to its defective turning responses (Heisenberg, Wonneberger, and Wolf, 1978). Males expressing *omb^{H31}* can see, but their optomotor behavior is dramatically reduced compared to a wild-type male's turning responses to a moving female (Tompkins et al., 1981). Also, there is a twofold drop in the measured courtship activity of *omb^{H31}* plus a decrease in its mating success (Tompkins, et al., 1981; cf. Cook, 1980).

Since an olfactory defect and a visual one each "removes" approximately half of the male's courtship activity, it was thought that a doubly mutant male might have all courtship eliminated, would be behaviorally sterile, and would support the suggestion that sight and smell are the only sensory modalities responsible for triggering reproductive behavior in this species. Indeed, mutant males simultaneously expressing the smell-blind and glass-eye mutations show almost no courtship of females (Tompkins, Hall, and Hall, 1980). However, given 10 days in the presence of females, these double mutants are able to mate in some cases; this suggests that some other modality, such as contact chemoreception or mechanoreception, is sufficient to allow mating under these dire circumstances.

The use of these single-gene mutations was critical in the above studies, especially in assessing the role of olfaction. The effects of vision can be tested by turning off lights, but it is very difficult to eliminate olfactory input into a fly, because receptors for such stimuli are apparently at several locations on the adult animal (Harris, 1972; Palka, Lawrence, and Hart, 1979).

Courtship between males of *D. melanogaster* usually occurs infrequently and not in a sustained fashion (Hall, 1978b; Jallon and Hotta, 1979); two males court one another with about the same measured vigor as is seen in the aforementioned tests of blind, olfactory-minus males with females. An exception to the low-level courtship between males is induced by the male-sterile fruitless mutation (see figure 24). These mutant males court each other or wild-type males; and wild-type males court them at abnormally high levels (Hall, 1978b). The volatile compounds from mutant males include materials that are superficially similar to those produced by wild-type females (Tompkins, Hall, and Hall, 1980). Thus, some of the aberrant courtship induced by *fru* may be due to the fact that the mutant male's pheromones stimulate him to court other males and stimulate other males to court him.

An anomalous aspect of the pheromonal control of courtship involving wild-type flies is seen in tests of very young adult males (for example, Cook and Cook, 1975; Jallon and Hotta, 1979). These males (6 hr old or less) stimulate much courtship as performed by mature males (3–7 days old), and the former, indeed, generate volatile compounds that their nonstimulating older brothers do not (Tompkins, Hall, and Hall, 1980). The age-specific compounds stimulate high levels of courtship between mature males in the bioassay discussed above (Tompkins, Hall, and Hall, 1980). Young *fru* males stimulate the same high-level courtship as do young wild-type males (additive effects of immaturity and the mutation are not seen); this suggests the possibility that the mutant males and the young males are generating similar or equivalent pheromones (Tompkins, Hall, and Hall, 1980). Therefore, one of the effects of *fru* may be simply to delay the turn-off of sex-stimulating pheromone production that usually occurs during the maturation of normal flies.

Why do young males stimulate courtship? Perhaps a mechanism has evolved for these immature flies to hear the courtship song in a sustained fashion. This information may be stored and contribute to the maturation of the male. The hours when immature adult males are exposed to mature song may be a critical period during which the young males store and use information that is relevant to the development of their full-fledged sexual behavior. Such a period would be analogous to the postnatal developmental stage that young birds experience, during which they must hear the mature complex song, performed by older males of their species, if they are later to develop the completely normal species-specific song, or at least the normal dialect, that corresponds to the one they have heard (for example, Thorpe, 1958; Konishi and Nottebohm, 1969; Marler and Peters, 1977).

The suggestion that learning and memory could, in part, mediate certain aspects of courtship comes from experiments involving mated females. In general, these flies will not accept a subsequent copulation for 4-10 days after mating (Manning, 1967b; Connolly and Cook, 1973). Such females also *stimulate* relatively low levels of male courtship activity (and very rarely copulate again). Siegel and Hall (1979) have shown that this poor stimulation obtains even for genetically blind males; thus rejection behaviors performed by females (Connolly and Cook, 1973) and received visually by males do not completely explain the low-level courtship. There is also poor stimulation by females that have been immobilized by either anesthesia or the effects of a temperature-sensitive paralytic mutation, compared to the reasonably good stimulation effected by immobile virgin females (Siegel and Hall, 1979; Tompkins et al., 1981; compare importance of movement detection as revealed by tests of the optomotor-blind mutant). It seemed, therefore, as if an aversive pheromone could be generated by mated females; and this has been shown to be so in tests of the different effects that volatile compounds from mated females have, compared to the more stimulating effects of such compounds from virgin females (Tompkins and Hall, 1981). The involvement of olfaction in the discrimination by males of virgin versus mated females is further shown by the fact that mutant smell-blind males cannot tell the difference between the two kinds of females and thus court them equally well (Hall et al., 1980c).

*Connections between Learning and Courtship in Drosophila*
The mated females induce an aftereffect on males that are placed with them for relatively long periods of time. These males—having been exposed to putatively inhibitory substances—show substantially reduced courtship of virgins for at least 2 hr after having been with the mated females. Such an aftereffect could be due to simple debilitation; but it could be that the males associate the virgin female with an

aversive stimulus and avoid her until the effects of the training have decayed. This hypothesis on conditioned courtship behavior is given some support in tests of the amnesiac mutant that was isolated by criteria of defective memory, not connected with defects in courtship (Quinn, Sziber, and Booker, 1979). Siegel and Hall (1979) found that amnesiac males sense the presence of mated females and court them at reduced levels. Their subsequent courtship of virgin females is also depressed. Yet the amensiac mutant males begin to respond well to virgins within 15–30 min after having been with the mated females, a much shorter time than the usual 2–3 hr required for wild-type males to court again normally. The use of the memory mutant in these courtship tests reveals that some features of reproductive behavior may not be solely in the realm of fixed-action patterns.

Tests of the dunce mutant in experiments with mated females showed not only that males of this genotype can discriminate between mated and virgin females but also that these males are normally conditioned to show poor courtship of virgins afterward (R. W. Siegel, unpublished). This result, though, does not disprove the involvement of learning in the mated female system because dunce has been found to learn normally in other tests as well, such as those involving color vision (Dudai and Bicker, 1978). Furthermore, preliminary tests on three other learning mutants, turnip, cabbage, and rutabaga (see section on "Learning and Memory") have shown that males expressing either of these X-linked mutations are poorly trained, if at all, by exposure to mated females (Hall et. al., 1980b).

**Neuroendocrine Features of Reproduction in Vertebrate Mutants**

In vertebrates, the sex-specific wiring and other aspects of morphology in certain parts of the CNS are not entirely under the control of the genotype of cells in the brain (for example, Raisman and Field, 1973; Gorski et al., 1978; reviewed by Arnold, 1980). This is because of the well-known hormonal control of sexual differentiation, including the development of the brain centers that are involved in reproduction, such as the critical *preoptic area* associated with the hypothalamus in mammals.

Some of the information on this hormonal control comes from studies of interesting genetic variants in vertebrates. One key variant is the X-linked *Tfm* mutation in the mouse (there is an analogous mutation in humans), which causes *Tfm/Y* animals to be phenotypically female, though, of course, sterile. The mutant phenotype essentially results from an absence of testosterone receptors, which must be present in the brain as well as other tissues for male differentiation (reviewed by Ohno, 1979). Indeed, in the wild-type male mammal, such receptors are present in the preoptic area of the hypothalamus, as are receptors for estradiol. *Tfm/Y* animals lack the testosterone receptors, but not those for estradiol. If the relevant receptors are present in these parts of the CNS that are critically involved in the control of reproduction, they can presumably respond to the hormones and thus influence the formation of patterns of synaptic connections formed in the proper sex-specific fashion (Raisman and Field, 1973). The intrinsic sex of these neurons is not relevant because administering male hormones to a female, or vice versa, at the correct developmental stages will cause the preoptic area to take on the synaptic characteristics associated with the administered substance (Raisman and Field, 1973), in addition to altering sexual characteristics in general, including courtship behavior. This nonautonomy of brain structure and function, in which the genotype of brain cells need not correlate with the structural or behavioral phenotype, is very different from the thoroughgoing autonomy that is seen in the control, by sex-chromosome genotype, of reproductive behavior in insects.

Another genetic variant relating to hormonal control of sexual development and behavior involves a diabetic (*db*) strain of mouse, which turns out to be neuroendocrine mutant. These mice are outwardly somewhat obese. For the current discussion, it is interesting that neither *db/db* females nor males show any mating behavior (Johnson and Sidman, 1979). The male reproductive system appears nearly normal. The mutant female exhibits no estrous cycle and, in fact, has an atrophied uterus. A key feature of the reproductive defect is that there is defective release of gonadotropin-releasing hormone (GnRH) from the hypothalamus, which often has abnormally high levels of GnRH in the diabetic mice. In these mutants, the pituitary can respond normally to exogenously administered GnRH (Johnson and Sidman, 1979). In further steps of the hormonal pathways, *db* is also apparently normal because the ovaries will respond to exogenous gonadotropin and the females will generally respond to administered estrogen. Thus, the focus for the reproductive defects may be in the hypothalmus, though the defect could actually be at the level of a failure to stimulate this tissue.

In the Mexican axolotl *Ambystoma* there is a mutant with reproductive characteristics analogous to those shown by the diabetic mutant. The eyeless mutant, with its several defects in embryonic induction and early brain development is also sterile. Yet the absence of the eyes does not affect fertility, as shown by Van Deusen (1973), who transplanted eyes into the mutant or removed eyes from developing wild-type salamanders. The latter treatment did not affect fertility, and there was no rescue of sterility by the former manipulation. Sterility associated with the mutant ovaries could, however, be corrected by putting these tissues into ovariectomized wild-type hosts. The suggestion, then, is that the eyeless mutation causes a lack of normal signals from the brain. Indeed, it has been found that, if the entire head of a normal embryo is transplanted onto a homozygous mutant body, a fertile adult results (Van Deusen, 1973). The reciprocal transplant leads to sterility (Van Deusen, 1973). The focus of the eyeless mutation with respect to sterility is likely to be the hypothalamus, which would not stimulate the pituitary to produce gonadotropin in mutants. This idea stems from the fact that the presumptive eye-forming region of the anterior embryonic ectoderm is very near the prospective hypothalamus. An additional experiment showed that a graft of normal hypothalamic primordium alone into mutant embryos led to fertile adults (see review by Epp, 1978).

Therefore, the eyeless mutation leads to a defect in a rather defined region of the ectoderm that, in the mutant embryos, may not respond to inductive signals, and the result is a pleiotropically abnormal phenotype. Yet the mosaic experiments, relating to defective eye development, sexual development, and other abnormalities as well, show that the pleiotropy has a comprehensive and unifying etiology, concerning essentially one defect in brain development and function. The same sanquine conclusion could be made in regard to the pleiotropy associated with the reproductively defective diabetic mouse. In general, genetic mosaics are a very important tool for determining whether a neurological mutant that also expresses other kinds of defects has one particular abnormality—which then gives rise to the others—or, instead, the relevant mutation has direct affects on several tissues simultaneously.

## NEUROCHEMISTRY OF HIGHER BEHAVIOR

A few genetic variants influencing behavior and putatively affecting neurotransmitter metabolism have been identified in mice and humans. These are largely uncharacterized in terms of the molecular nature of the genetic lesion and are even rather poorly characterized as to such basic properties as mode of inheritance or map position. However, the dramatic behavioral attributes associated with these heritable conditions and the neurochemical correlates that appear to accompany them endow them with particular significance.

*Catecholamine Variants in Mouse*
In the mouse, inbred strains have been found to exhibit markedly different behaviors with respect to aggression (reviewed by Lagerspatz and Lagerspatz, 1974). These strains may also have significantly dif-

ferent levels of enzymes required for catecholamine synthesis: tyrosine hydroxylase, dopamine-$\beta$-hydroxylase, and phenylethanolamine $N$-methyltransferase. One such "aggressive" strain has elevated levels of all three enzymes (Ciaranello, Lipsky, and Axelrod, 1974). Transmission studies indicate that the aggressive behavior segregates in a mendelian fashion as an autosomal recessive locus. The segregation characteristics of the enzyme phenotype unfortunately were not reported. Mapping data collected simultaneously on both phenotypes are required before it can be forcefully argued that one mutant gene gives rise to both the biochemical and behavioral changes.

*Human Psychoses and Other Biochemical Neuropathies*
The number of behavioral and neurobiological abnormalities in humans that are known or suspected to have a genetic etiology is legion. We shall not attempt to cover or even mention any but a small fraction of these cases (see also reviews by Childs, 1972; Childs et al., 1976). Instead, we shall focus on some very selected cases, which in general have to do with "higher behaviors" in this organism, thus excluding, for instance, cases of gross and general retardation that can arise genetically (for example, due to trisomy of the autosome, chromosome 21). The abnormalities, or presumed abnormalities, of higher functions to be considered will be cases in which the evidence for an authentic genetic etiology that is at least partially responsible for the aberrant behavior is solid. Certain of these behavioral problems are also associated with some information that is neurochemical and are not limited strictly to behavioral findings. Thus, these cases can well be placed under the heading of "neurogenetics," as opposed to the vast array of behavioral syndromes or abnormalities in humans, so many which have no known biological concomitant, and nearly all of which are purported to be due to polygenic genetic differences or an "autosomal dominant mutation with variable penetrance."

**Schizophrenia**

Schizophrenia is a disease of the brain affecting a spectrum of neurological functions. It is usually accompanied by disorders in the sensory system, such as auditory hallucinations; disorders in the motor system, such as a muteness or rigidity; and disorders in consciousness such as perplexity or disorientation, together with impaired memory and attention. Schizophrenia seems to be caused by a mixture of genetic and environmental factors. The following points summarize some of the evidence implicating the genetic transmission of the disease: (1) The risk of schizophrenia for any person increases with his or her degree of relatedness to a schizophrenic relative. (2) Identical twin concordance rates (50%) are about three times those of fraternal twins. (3) Adopted children of schizophrenics have a much higher risk of the disease than adopted children of normal parents. (4) Adopted relatives of schizophrenic adoptees do not have elevated rates of schizophrenia, but biological relatives of these adoptees do. (5) Identical twins reared apart are about as concordant for schizophrenia as those reared together (Gottesman, 1978; Kety et al., 1978).

Among the environmental factors that seem to increase the risk of schizophrenia are the following: (1) postnatal brain damage, (2) season of birth (January to April), and (3) sociofamilial inadequacies. Interestingly, these environmental factors seem to be correlated with schizophrenia only among those who are known to have a low genetic risk for the disease (reviewed by Kinney and Jacobsen, 1978). There must be other factors besides these that influence the probability of developing schizophrenia among those genetically predisposed to the disease; otherwise, concordance rates for monozygotic twins would be close to 100% instead of half that. These additional factors are not yet known.

The best predictor of schizophrenia at present, then, is the genetic one. By analyzing the data, one can construct specific models of genetic transmission. One model that has been proposed, though it is probably oversimplified, suggests that schizophrenia may be due to a variant at a single autosomal locus (reviewed by Kidd, 1978). Homozygotes for this allele would have about a 50% chance of developing the disease, while heterozygotes would have a 25% chance. More complicated polygenic models can, of course, explain the data just as well, and at present there seems to be no basis for selecting the most appropriate model. There is the hope, however, that neurochemical correlates of schizophrenia will shed new light on the genetics.

A genetic etiology of schizophrenia necessitates a biochemical explanation of the disease, and a variety of evidence now suggests that the metabolism of the CNS transmitter dopamine may be primarily affected in schizophrenics (reviewed by Snyder, 1978). The first piece of evidence linking dopamine metabolism and schizophrenia is that neuroleptic drugs, or sedatives, such as phenothiazines and butyrophenones, that have profound antischizophrenic actions have been shown physiologically and pharmacologically to block dopamine receptors (Snyder, 1978). There is, in fact, a good correlation between the affinity of such an agent for binding to the dopamine receptor and its clinical potency. Reserpine, which works through a completely different pharmacology to reduce dopamine neurotransmission, also alleviates schizophrenia. Finally, $\alpha$-methyl-$\rho$-tyrosine, which blocks dopamine synthesis, potentiates the effects of other antipsychotic agents (Snyder, 1978).

The next most important evidence supporting the dopamine theory of schizophrenia is that a wide variety of antidepressants, such as amphetamine, methylphenidate, or cocaine, which act by releasing dopamine, potentiating dopamine action, or blocking dopamine reuptake, can, at low doses, produce tremendous exacerbations of schizophrenic syndromes when given to active schizophrenics. These drugs worsen preexisting symptoms; hebephrenics become more hebephrenic, catatonics more catatonic. Furthermore, when given to normal subjects at high doses, these drugs can cause paranoid schizophrenia very similar to the clinical syndrome (Snyder, 1978).

The dopamine theory of schizophrenia has also been supported by reports of significantly decreased levels of monoamine oxidase (MAO), an enzyme that in the brain normally breaks down released dopamine, in the platelets of schizophrenic patients. Other enzymes involved in dopamine metabolism are also under investigation in schizophrenics (reviewed by Barchas, Elliott, and Berger, 1978).

Some dopaminergic pathways in the brain terminate in the limbic system, which largely regulates emotional behavior; others go to parts of the frontal, cingulate, and entorhinal cortex, linked in function to the limbic system but poorly understood. Lesions in the frontal cortex can lead to subtle or profound changes in personality. It would certainly not be surprising if schizophrenia involved abnormalities in these parts of the brain.

The dopamine theory of schizophrenia has far to go before it can be totally accepted, and there are some conflicting pieces of evidence. Thus, if not wrong, the theory is probably oversimplified. Furthermore, it is not at all clear whether changes in dopamine metabolism in schizophrenics, if they exist, are causal or secondary. Nevertheless, the dopamine theory is a useful and stimulating hypothesis connecting the genetic etiology of schizophrenia with the mental disorders.

## Affective Disorders

Certain major affective disorders in humans also give evidence of heritability and associated neurochemical changes. The presence of a genetic component in these illnesses is strongly suggested by the high incidence of the disorders in first-degree relatives of patients and the higher rate of simultaneous illness in monozygotic twins than in dizygotic twins (reviewed by Sachar and Baron, 1979). However, the mode of inheritance is not at all clear; nor is it established that the two major conditions, recurrent (unipolar) or manic (bipolar) depression, are different forms of the same illness with shared genetic determinants (reviewed by Gershon et al., 1977).

There have been many studies of enzyme changes associated with affective disorders, but most of the findings have been inconsistent or inconclusive (Gershon et al., 1977). One alteration in enzyme activity that does correlate with the presence or absence of illness in families showing a high incidence of affective disorders is that of catechol-$O$-methyltransferase (Gershon and Jonas, 1975). This enzyme, thought to be involved in catecholamine degradation in the brain, is present in erythrocytes, where its levels can be conventiently measured. Higher levels of it are found in the red blood cells of healthy subjects than in their sick relatives (Gershon and Jonas, 1975). However, there is no certainty that erythrocyte levels are an accurate reflection of brain levels (Gershon et al., 1977). There have been many reports of altered levels of metabolites of catecholamines in urinary excretion and cerebrospinal fluids in these patients (Gershon et al., 1977). Although no conclusive data for any correlation with affective disorders exist, there is some evidence for persistent reductions in the level of a metabolite or norepinephrine in the cerebrospinal fluid of these patients (Goodwin and Post, 1975).

**Lesch-Nyhan Syndrome**

The Lesch-Nyhan syndrome in humans is a genetic disease that is X-linked and recessive (Nyhan, 1973). The first signs of the disease, which usually develop at 6–8 months of age, consist of crystalluria, hematuria, and kidney stones. Cerebral manifestations occur shortly thereafter. Sitting and holding up the head becomes difficult; choreic and athetoid movements appear with severe spasticity. Usually there is mental retardation. The most severe and bizarre aspect of the syndrome is a compulsive self-destruction. Usually the lips and fingers of Lesch-Nyhan children are excessively mutilated from self-inflicted bites. Since these patients do not have sensory deficits, they scream in pain when they bite. This aggression is not exclusively self-directed, in that patients will also attack others, although usually with small success owing to motor problems. As the patient grows older, he also becomes aggressive in speech.

The genetic defect in Lesch-Nyhan's disease is understood at the molecular level as a deficiency of the enzyme hypoxanthine-guanine phosphoribosyltransferase (HGPRT) (Seegmiller, Rosenbloom, and Kelley, 1967). It is comprehensible why such a deficit can lead to the hyperuricemia associated with the disease because this enzyme is important in the feedback control of purine synthesis in man. Uric acid is the end product of purine catabolism; hence, there is an over-production of purines and uric acid. The basis of the behavioral abnormalities, however, is considerably more of a mystery.

Some findings suggest that catecholamine metabolism may be altered in this disease. First, in animals the administration of dopamine (Costall and Naylor, 1975), the dopamine agonist apomor-phine (Ernst, 1967), or the $\alpha$-adrenergic agonist clonidine (Razzak, Fujiwara, and Veki, 1975) can induce some of the neurological signs, such as choreic movements and self-mutilation. Second, patients display lower than normal increments in plasma norepinephrine levels in response to stress (Lake and Ziegler, 1977). Finally, cultured Lesch-Nyhan cells or any cells deficient in HGPRT routinely show low levels of monoamine oxidase activity (reviewed by Breakefield, Edelstein, and Castro Costa, 1979). MAO is important in the breakdown of cate-cholamine neurotransmitters.

Another possibility is that glycine metabolism is affected in the disease. Glycine, almost certainly a neurotransmitter in the mam-malian CNS, is present in abnormally high intracellular concentra-tions in HGPRT-deficient mutants of glioma and neuroblastoma cells (Breakefield, Edelstein, and Castro Costa, 1979).

A still vaguer possibility is that variations in adenine metabolism may account for the neural defects in Lesch-Nyhan patients. Adenine and its analogues, AMP, ATP, cAMP, and so forth, are certainly of great importance in the proper functioning of neurons and synapses. Fibroblasts from Lesch-Nyhan patients require adenine for growth, unlike control fibroblasts (Felix and DeMars, 1969). The activity of adenine phosphoribosyltransferase is increased in Lesch-Nyhan patients, and phosphodiesterase activity is increased in neuroblastoma lines lacking HGPRT (Seegmiller, 1976).

It is not understood how the above defects in metabolism that appear to be subsidiary to the HGPRT deficit result from the primary enzymatic abnormality. It seems likely that, whatever the neuro-chemical explanation of the disease turns out to be, the best candidate for at least part of the behavioral dysfunction is the area comprising the basal ganglia. The basal ganglia contain very high levels of HGPRT in normal brains (Rosenbloom et al., 1967). These ganglia are also known to be involved in other diseases causing choreiform movements, such as Huntington's disease and Fahr's disease. It is not known what affected brain areas might be associated with the aggres-sive self-mutilating behavior. Lesions of the basal ganglia do not seem to release such behaviors.

In short, much remains to be learned about the neurobiological basis of this disease. Among the questions to be resolved are the following: What chemicals connect a deficiency in HGPRT activity with the behavioral changes? What is the causal chain of events? What parts of the brain are involved?

# 4
# DEVELOPMENTAL
# NEUROGENETICS

## FATE MAPPING OF PRESUMPTIVE NEURAL TISSUES

The neurogenetic applications of genetic mosaics in *Drosophila* have already been exemplified several times (see Palka, 1979a, for review). One of the principal additional uses of these mosaics has been to generate developmental fate maps of early *Drosophila* embryos. These fate maps have been, in turn, applied to studies of parts of the fly affected by many of the behavioral mutants already discussed (Hall, 1978a). It was hoped that the "mapping of behaviors" would provide information on parts of the early embryo that develop into excitable cells and tissues, especially internal ones that may not be easily accessible to analysis.

*Drosophila* mosaics used in fate mapping arise by induced loss of one X-chromosome in early embryonic nuclear division in an embryo that began as a diplo-X, female zygote (figure 30; of figure 15). Such chromosomal loss can be effected through the use of "unstable" X-chromosomes (for example, ring chromosomes), or regular X-chromosomes that have been subjected to the effect of mutations in *Drosophila,* which lead to unstable chromosome behavior during meiotic and mitotic nuclear and cell divisions (see Hall and coworkers, 1976, for a review of these mosaic-generating techniques). It is useful to keep in mind that in fate mapping and for the mosaic analysis of neurological mutants, the heavily used diplo-X, haplo-X gynandromorphs are not the only possible approaches. These sex mosaics are the most readily obtainable, however, since the cells of these organisms tolerate the haplo-X condition; somatic loss during development of any of the other chromosomes (autosomes) leads to poor viability or lethality of the zygote.

The ground plan of the embryo revealed by fate mapping includes the locations of progenitor cells (landmarks) for the sensory structures and central ganglia of *Drosophila* that have been discussed throughout this book. Janning (1978) has summarized the principles of, and results on, most of the recent mosaic fate maps, including a discussion of internal markers used to place presumptive neural ganglia on the map. The use of these cellular markers—involving enzymatic functions of neurobiological interest (for example, acetylcholinesterase) or others that are of interest only as markers (for example, acid phos-

FIRST MITOTIC DIVISION

XX (FEMALE)                    X (MALE)

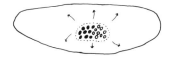

NUCLEAR DIVISIONS

BLASTODERM

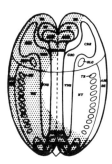

FATE MAP OF
BLASTODERM

LARVA

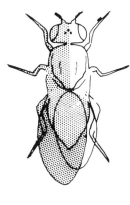

ADULT

phatase)—is an important recent advance in mosaic fate-mapping technology. Another is the construction of the maps by rapid, computer-aided procedures (Flanagan, 1976). These automated techniques are more valuable than mere time-saving devices, for, in addition, they achieve the best landmark positioning, including an error analysis, in an objective fashion. Examples of computer-constructed fate maps, including data from scoring genotypes of tissues in the CNS, are shown in figures 31 and 32. The map in figure 31 refers to the placements on the embryonic blastoderm of progenitor cells for tissues in the adult fly (Kankel and Hall, 1976), principally the embryonic locations of primordia for the neural ganglia and the imaginal disks. The latter are tissues set aside during embryonic development, destined to grow but not differentiate during larval development, and then finally metamorphose during the pupal period into structures of the adult cuticle. All the sensory cells and axons of the adult are derivatives of the imaginal disks.

The other fate map (figure 32) shows the landmarks for the larva, that is, positions on the blastoderm of primordia for that intermediate stage of the life cycle. This map (D. R. Kankel, unpublished) is not wholly consistent with the map for adult precursors, for example, in terms of the relative embryonic distances between neural and imaginal disk landmarks. However, the mosaic analysis of these internally marked larvae does reveal that cells in the larval CNS per se (that is,

**Figure 30.**
Origin and development of gynandromorph in *Drosophila*. The fertilized egg in this diplo-X female carries one X-chromosome in the shape of a ring (arrow); this X has normal, wild-type alleles of all marker and behavior genes on the chromosome. The other X is in the normal rod shape and carries mutant alleles of marker genes (such as yellow body color and white eye color) and a behavior mutation if determination of the focus of this gene action is to be studied. The ring-X behaves unstably in early mitotic divisions in the zygote, so that 10–30% of these female embryos will lose this X during such a division and become a gynandromorph with female nuclei (solid circles) and male nuclei (open circles). The resultant cells and tissues in this mosaic will be part female (shaded) and part male. The distribution of tissues of different sexual genotype and phenotype is different in separate gyandromorphs (for example, anterior female and posterior male, instead of the left-right division shown here). This creates, among several mosaics, a variety of "dividing lines" between unmarked (female) and marked (male) cuticular tissues that reflect dividing lines separating normal from mutant (marked) progenitor cells (at the blastoderm stage. If two progenitor cells are relatively far from each other, dividing lines will commonly pass between their locations, and, hence, the tissues eventually developing from these embryonic cells will frequently be of opposite genotype and phenotype among the several mosaics analyzed. Conversely, structures deriving from nearby cells at the blastoderm stage will be, with relative infrequency, of opposite genotype in these mosaics. The higher frequencies are converted into relatively long fate-map distances (for example, between sites in the thoracic ganglia and PRO, the site for the proboscis, in figure 31). Lower frequencies become relatively short distances (for example, between the blastoderm site for the eye and that for OPTLB, the optic lobes, in figure 32). [Benzer, unpublished]

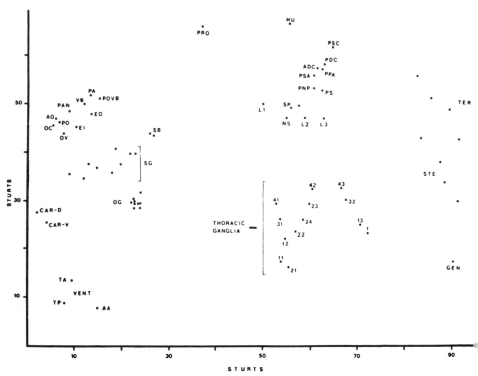

**Figure 31.**
Fate map of the embryonic blastoderm of *D. melanogaster:* progenitors of adult. The
map was constructed with the least-squares procedure of Flanagan (1976), using data
from the scoring of genotypes of external and internal structures of adult gynandromorphs (Kankel and Hall, 1976). STURTS on the axes are units of fate-map distances
(Hotta and Benzer, 1972). Explanation of symbols (see Kankel and Hall, 1976, and
Flanagan, 1977, for additional details): group including AO, PA, EI, bristles, sensory
structures, and appendages of external head; SB, sites in cortex of ventral brain
(subesophageal ganglion); SP, sites in cortex of dorsal brain (supraesophageal
ganglion); OG, sites in cortex of optic lobes (ganglia); two CAR sites, parts of valve
(cardia) in thoracic gut; group surrounding VENT, parts of thoracic and abdominal gut
(ventriculus); PRO, proboscis; HU, bristles of humerus on dorsal thorax; group including PSC, PNP, L1, NS, bristles and appendages of thorax; thoracic ganglia, sites in cortex of ventral nervous system (numbered as in Kankel and Hall, 1976); TER, dorsal
parts (tergites) of abdominal segments; STE, ventral parts (sternites) of abdominal segments; GEN, external genitalia. [Flanagan, 1977]

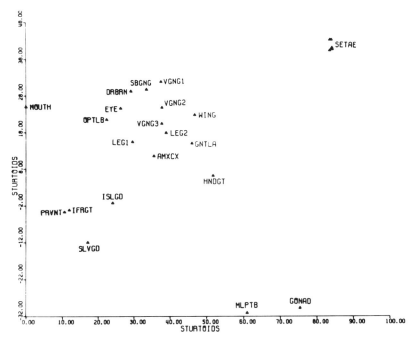

**Figure 32.**
Fate map of the embryonic blastoderm of *D. melanogaster:* progenitors of larva. The map was constructed with the least-squares procedure of D. R. Kankel (unpublished), modified from Flanagan (1976). The genotypes of various tissues in larval gynandromorphs were scored using a cuticular marker (yellow) or sexual dimorphism (regarding the GONAD); genotypes of internal tissues were scored using the acid phosphatase marker (cf. Kankel and Hall, 1976). STURTOIDS on the axes are units of fate-map distances (reviewed by Hall, Gelbart, and Kankel, 1976). Explanation of symbols: MOUTH, larval mouthhooks; OPTLB, formation centers in larval brain for optic lobes of adult; EYE, imaginal disk for eye; DRBRN and SBGNG, sites in dorsal and ventral brain, respectively, of larva; VGNV1–3, sites in ventral nervous system of larva (near inserts of thoracic imaginal disks); LEG1, LEG2, WING, thoracic imaginal disks; SETAE, ventral "feet" of abdominal segments; AMXCX, anterior sensory complex; HNDGT, larval hindgut; MLPTB, malpighian tubules; ISLGD, presumptive salivary gland of adult; SLVGD, larval salivary gland; IFRGT, presumptive foregut of adult; PRVNT, larval foregut. [D. R. Kankel, in Hall and Greenspan, 1979]

formed during embryogenesis) are very closely related to precursors of the adult CNS. Thus, the functioning larval cells are virtually always of the same genotype as nearby neuroblasts that will become parts of the CNS specific to the adult, for example, the optic lobes.

Mosaic fate mapping has not produced a great deal of fundamentally important neurogenetic information, in spite of the recent placements of neural progenitors (for example, figure 31). Findings from the mosaics as useful as the map constructions are the data that have allowed estimates of the number of blastoderm cells giving rise to each major ganglion. The numbers of such progenitor cells are small (approximately 5–20 per ganglion). These conclusions are the result of calculating the frequency with which a ganglion is of mixed genotype, or, alternatively, the smallest fraction of a ganglion that is one of the two relevant genotypes in a given mosaic. That the number of progenitor cells for a given ganglion is neither one cell nor a huge number is consistent with the patterns of mosaicism seen with respect to external parts of the fly, many of which develop from the imaginal disks (Merriam, 1978). These data on gynandromorphs generated early in development imply that, for the fly's nervous system, it is difficult to produce such mosaics that will end up with a tiny fraction of a ganglion of one genotype. It would be desirable to analyze behaviorally such small groups of mutant (or wild-type) cells in mosaic experiments on neural "foci" of behavioral variants.

The problems involved with using mosaics to determine the anatomical site at which the mutant gene exerts its primary effect (that is the focus) have not deterred investigators from applying these techniques. Developmental, physiological, and behavioral mutants have been morphogenetically mapped in mosaics (Hall, 1978a), and almost all these studies have involved abstract fate mapping of the mutational defects, that is, a placement of the foci on the maps by triangulation from cuticular landmarks (see Benzer, 1973). Thus, there has been no direct demonstration of the tissues affected in these cases. Sometimes, though, the mosaic experiments have succeeded in ruling out extrinsic influences on the tissues presumed to be directly affected by neurological mutations. For instance, blind mutants such as *norpA* (see "Sensory Mechanisms") were shown to be autonomous to the eye. If, for example, one eye of a fly is genotypically mutant and the rest of the fly normal, then the mutant eye does not have its functional defect "corrected" by circulating material that hypothetically would have been produced by the normal tissues (Hotta and Benzer, 1970). This is not just a "straw man," for there is one case in which abnormal vision was found to be due to a primary defect elsewhere in the animal. This is the putative genetic disease in humans that leads to pigmentary degeneration of the retina. It was eventually found to be caused by faulty absorption of vitamin A by the gut (Gouras, Carr, and Gunkel, 1971). Since this disease also has pathologies besides those in the eye (for example, Fredrickson, Gotto, and Levy, 1972), a nonretinal focus is not surprising. These findings were not made with mosaics, and, indeed, the tests of *norpA*'s autonomy did not require the application of complex fate-mapping algorithms (see Hall, 1978a; Janning, 1978), but merely the use of an eye-color marker that tagged a given eye in a mosaic as normal or mutant.

When the detailed algebraic and statistical procedures of mosaic analysis are applied, there are several computational difficulties associated with the mapping of neural foci, especially those related to complex behaviors performed by the whole animal (as opposed to a defect in an individual sensory structure or appendage). In addition to these complexities, the general locations of the whole-animal behavioral foci are often not interpretable as to what part of the embryonic blastoderm they refer to. The great majority of these foci have mapped to "no man's land" on the ventral, anterior portion of the blastoderm that could be presumptive neural, muscular, or even alimentary tissue (Hall and Greenspan, 1979). Another problem is that the statistical errors in the placement of foci are so large (for example, Flanagan, 1977) that if two foci, from separate studies, appear in different parts or the same part of their respective fate maps, it is not valid to conclude that the sites of the mutational defects are in different tissues or the same tissues, respectively. Therefore, overlaying different behavioral maps may not be a very useful procedure; nor would an overlaying of behavioral maps with the straightforward fate maps that include the positions on the blastoderm of presumptive internal tissues (for example, figure 31).

The major usefulness of the analysis of behavioral mutants in mosaics may be that it has sometimes stimulated study of tissues presumed to correspond to the focus. For example, Hotta and Benzer (1972) mapped the focus, a putatively neural one, for the drop-dead (*drd*) mutation before finding that *drd* does cause brain degeneration in adults expressing this late-acting lethal mutation (figure 33). However, if they had sectioned *drd* flies without having previously collected mosaic data, they would still have seen the brain degeneration because it is gross: a "spongy" appearance in all the major ganglia that is not dissimilar from the neural degeneration seen in very old adults of *D. melanogaster* (Miquel, 1972).

The mosaic analysis of *drd* does not prove a primary neural focus for the gene action because internal marking of *drd* and *drd*⁺ cells has not been carried out. Thus, it is not rigorously correct to conclude that the focus of *drd* maps the location on the blastoderm of progenitors of the brain. The major piece of useful information derived from the

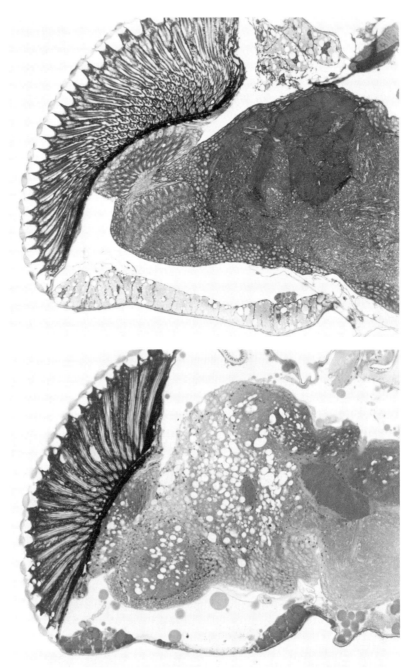

**Figure 33.**
Horizontal sections through head of *Drosophila,* showing eye (at extreme left), optic ganglia (at left of center), and brain. (Top) Normal fly. (Bottom) drop-dead mutant, at stage of pronounced staggering (toluidine blue); × 75. [Benzer, 1971]

mosaic experiment is that *drd*'s defect appears to require mutant tissue on both sides of the animal (that is, a "submissive" focus; Hotta and Benzer, 1972).

It is possible, however, that a putatively neurological mutant will have no overt anatomical abnormality. For example, a mosaic study of the doomed (*dmd*) mutation (like *drd,* a late-acting lethal) is of some use, aside from the fact that the objective computer-aided mapping techniques were employed. Additionally, Flanagan (1977) found a midventral location for the lethal focus (cf. figure 31); this is the only hint that there may be a neural focus for the action of *dmd,* since no morphological change in the CNS occurs in *dmd* flies (Flanagan, 1977).

Can mosaic studies of neurological mutants be used in powerful ways? We suggest that some of the experiments, to be discussed, on interactions among excitable cells in mosaics provide unique and valuable information. Sometimes, in addition, the tissue mapping of mutational defects leads to rather specific information on the effects of neural abnormalities. A recent example is provided by the mosaic analysis of lethality induced by the aforementioned acetylcholinesterase mutations in *Drosophila* (see mosaic procedure in figure 15). Some other lethal mutants have been tentatively suggested to have neural foci by the abstract approach to mosaic fate mapping (for example, Bryant and Zornetzer, 1973). The approach to mapping *Ace*-induced lethality was more straightforward because *Ace*⁺ and *Ace*⁻ portions of the nervous system in a mosaic that had developed to adulthood could be detected by direct observation of stained and unstained tissues (Greenspan, Finn, and Hall, 1980). It was found that not only is the neural expression of this gene relevant but a particular portion of the CNS was most sensitive to the *Ace*⁻ genotype. This means that extremely few mosaics recoverable as adults had clones of mutant tissue—even small ones—in the posterior midbrain in the left *or* right side of the head. Unilaterally *Ace*⁻ clones elsewhere (for example, dorsal brain, thoracic nervous system) were found in much higher frequency and sometimes were large. Bilaterally mutant clones, simultaneously mutant in left and homologous right portions of the CNS, were seen in reasonable frequency in some parts of the brain, especially the optic lobes. But no bilaterally mutant clones were ever recovered in the posterior midbrain. Therefore, although this nervous-system-specific enzyme is present throughout the CNS in *Drosophila,* its activity is not crucial for viability in a spatially uniform fashion.

Control mosaics of the nervous system, with ganglia or parts of them marked with a nonlethal mutation such as acid-phosphatase-minus (Kankel and Hall, 1976; Greenspan, Finn, and Hall, 1980), have been routinely found to have large unilaterally mutant clones, small and large bilaterally mutant ones, and also to have a virtually *uniform* spatial distribution for the occurrence of clones anywhere in the brain or thoracic ganglia. Thus, the "experimental" mosaics for *Ace* show that there is an essentially lethal focus, due specifically to the absence of AChE activity in a particular region of the brain. This could not have been revealed by purely external mapping of the *Ace* mosaics followed by algebraic treatments of the data (Janning, 1978), no matter how sophisticated or automated (Flanagan, 1977).

It is also worth mentioning that large *Ace* clones are not only extremely rare in the mosaics for this mutation, but also appear to degenerate (see "Physiological Activity in Development"). Thus, so-called developmental lethality induced by *Ace* could be due to rather late defects in the maintenance of normal neural structures that are probably relevant to the defects induced by the drop-dead and doomed mutations as well. This suggests that neurogenetic issues in the developmental area do not cease to be important when the organism reaches the adult stage. Questions related to genes that "maintain" excitable tissues once they have developed will also appear later in a discussion of muscle differentiation.

## HISTOGENESIS AND DIFFERENTIATION OF EXCITABLE CELLS

### Neural Mutations Affecting Embryogenesis

It may be thought that mutations that affect the histogenesis of nerve cells, especially during early stages of development, will have such severe and widespread repercussions as to preclude survival. There-fore, these mutations would be lethal and consequently difficult or impossible to maintain in genetic stocks and thus unavailable for detailed analysis. This notion is borne out to some degree by examining the array of mouse mutations that affect brain development; mutants such as reeler and weaver (see later sections) do exhibit many of their abnormalities in postnatal stages, not in very early histogenesis and cell differentiation. These mutations are not lethal, having been identified by defective behavior in neonates (for example, Caviness and Rakic, 1978; Sidman, Cowen, and Eicher, 1979).

Analyzing early neural development using mutations is not a problem in *Drosphila*. Many lethal mutations of this organism have been isolated, and they are easily maintained in stocks, in heterozygous condition with marked, inverted "balancer" chromosomes. Several mutants affect the nervous system, even at very early stages of histogenesis. These mutants have been analyzed in three types of study: (1) examination of uniformly mutant embryos, segregating normal embryos from heterozygous parents; (2) examination of genetic mosaics that can survive and exhibit the effects of the lethal mutation in certain parts of the organism, including parts of the CNS; and (3) manipulation of conditional mutants, involving temperature-sensitive (ts) alleles of genes affecting histogenesis that can be "turned off" at a variety of stages of the life cycle, including very early ones.

Of the small number of lethal mutants that have been analyzed in terms of possible effects on the nervous system, some of the most important are alleles of the Notch gene. Notch, on the X-chromosome, is a putative "complex" locus or gene cluster (reviewed by Welshons, 1965; Wright, 1970), and it can be lethal in many of its mutant forms. Poulson (cited in Wright, 1970) noted that an early defect in the development of embryos hemizygous for Notch was an increase in the numbers of neural precursor cells, such that the early embryo forms an enlarged nervous system. What is also striking about these embryos is that the increased neural histogenesis appears to have occurred at the expense of hypodermal histogenesis (compare mutants in nematodes discussed later). Thus, the effects of Notch can be interpreted as a takeover by presumptive nerve cells (figure 34) of a portion of the embryonic fate map that is usually destined to form hypoderm (and thus eventually the larval cuticle). This fate map was determined by Poulson (1950), not with the mosaic techniques discussed in the previous section, but by direct observation of embryonic cells. Specifically, the model shown in figure 34 could mean that cells that usually enter the hypodermal pathway as they undergo their early divisions and determinative processes are "switched" into a neural pathway. It seems, then, that Notch causes the histogenesis of these tissues to go awry in a more interesting and unexpected fashion than would be the case if there were a simple absence of the histogenesis of early nerve cells.

One additional feature of the neuropathology associated with Notch mutants is that anterior tissues from mutant embryos, transplanted into adult hosts, form a lethal, invasive growth that is "teratoma-like" (Gateff and Schneiderman, 1974). The effects of this mutation on nerve cells is thus analogous to the neoplastic effects on such cells of another mutation, lethal-(2)-giant-larva (reviewed by Gateff, 1978). Both of these genes may be involved in the control of the size of the neuronal polulation, Notch during the embryonic stage and lethal-(2)-giant-larva in the postembryonic period.

Another X-linked mutation, similar in some of its effects to $N^{ts}$ mutants, is shibire-temperature-sensitive ($shi^{ts}$), which, as noted earlier, was discovered as an adult paralytic mutant (Grigliatti et al., 1973). Heat treatments in earlier stages induced a variety of developmental abnormalities (Poodry, Hall, and Suzuki, 1973), and the pat-

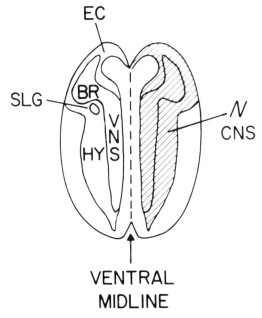

**Figure 34.**
Embryonic fate map of *D. melanogaster,* showing effects of the Notch lethal mutation. This ground plan of the embryo was determined by Poulson (1950). The embryo is split open at the dorsal midline and viewed from above. The left half of the fate map shows regions that normally develop into the larval brain (BR), ventral nervous system (VNS), hypoderm (HY), "ectoderm" (EC), and salivary gland (SLG). The right side shows the phenotype of the Notch (*N*) mutant (after Wright, 1970), in which the diagonal lines indicate portions of the embryo that develop into neural tissues (*N* CNS). In the mutant, the presumptive neural area is abnormally large on both the right side (as shown) and left side of the embryo. [After Poulson, in Wright, 1970]

tern of defects resembles those seen after heat treatments of a developing $N^{ts}$ mutant (Shellenbarger and Mohler, 1978). Also, Buzin and coworkers (1978) have shown that the differentiation of embryonic muscles and nerves in cultures of $shi^{ts}$ cells is inhibited by heat treatment. At a still earlier stage of development, $shi^{ts}$ induces defects in the formation of pole-cell membranes (that is, presumptive germ cells; Swanson and Poodry, 1980). It would be interesting to examine the effects of other lethal mutations at these cellular levels, such as Notch or the other early neural lethals to be discussed.

There is at least one other gene on the X-chromosome of *D. melanogaster,* affecting early neural development, whose action may influence similar features of histogenesis that are controlled by Notch. The lethal-X-20 (*l(1)X20*) mutation leads to developmental abnormalities of the nervous system, such that hemizygous mutant embryos exhibit three different kinds of developmental abnormalities (Ede,

1956). The majority express a neural defect similar to that caused by Notch (which maps near *l(1)X20,* but is still well over 100 genes away). A small proportion of the mutant embryos (about 10%) expresses a lethal abnormality that is the complementary phenotype of the Notch-like defect; these embryos (which can come from the same parents that generated the other kinds of defective embryos) have virtually no ventral neuroblasts and appear to have an enlargement of hypodermal anlagen. The remainder of the lethal embryos exhibit little or no early development, arresting at or before the blastoderm stage. Detailed genetic analysis of the *l(1)X20* genotype was not performed; so it was never proved that all of these mutant phenotypes were due to a single-gene mutation rather than to a more complicated genetic change.

Assuming for the sake of argument, though, that *l(1)X20* was a single-gene mutation (the strain was eventually lost), the complementarity of the first two phenotypes suggests a mechanism such as the "bistable switches" that have been invoked to understand pathways of determination and differentiation of the imaginal disks in *Drosophila* (Kauffman, Shymko, and Trabert, 1978). Thus, in the histogenesis of the embryonic ectoderm in wild type, some cells would be switched into a neural pathway, others into a hypodermal one (see figure 34). In Notch, a stable but incorrect switching of all these particular cells into the neural pathway can be invoked. In *l(1)X20,* different embryos may have an abnormal and labile switching event, sometimes sending all the relevant ectodermal cells into the neural pathway, other times failing to produce the majority of nerve-cell progenitors. These switching events, moreover, would appear to occur at the level of the organism because a given embryo expression *l(1)X20* is not a mixture of areas with too little neural development and others with too much; rather, the entire animal has gone completely into one pathway or the other.

While these abstract considerations are useful perhaps only as a mnemonic device, they may be testable due to some recent work. First, García-Bellido and Santamaria (1978; García-Bellido, 1979) have performed genetic and some developmental characterization of lethal variants— especially small deletions—at and very near the scute locus on the X-chromosome of *D. melanogaster.* This is the approximate position to which *l(1)X20* had been mapped. These findings suggested that certain of these deletions may derange embryonic development of the nervous system (García-Bellido and Santamaria, 1978). Such suggestions were followed by the more direct analysis of Jiménez and Campos-Ortega (1979; Campos-Ortega and Jiménez, 1980) and White (1980), who examined internal tissues affected by several deletions and have observed defects such as grossly reduced amounts or disorganizations of embryonic neural tissues. In addition, White (1980) has discovered that an apparent single-gene mutation—mapping to one subsegment of the X-chromosome in and around the scute locus— leads to embryonic lethality. More specifically, the mutation causes striking defects in the early development of the ventral nervous system and is thus called *vnd* (ventral-nervous-system-defective). The basic phenotypic defect, observable relatively late in embryonic development, is a failure of the ventral nervous system to condense normally (that is, become "cephalized" in the anterior portions of the animal). Part of the anterior developing gut is also abnormal in *vnd* embryos, but there are no other obvious defects. Thus, the action of this gene is relatively specific to the nervous system. The discovery of this mutant is important, first because it has very similar defects in development to those caused by the very small genetic deletion that takes out this genetic locus and a few neighboring genes (White, 1980). This seems to suggest that a search for regions of the *Drosophila* genome that result in neural defects when deleted will lead to the identification of individual genes controlling histogenesis, early cell differentiation, and early pattern formation in the nervous system (Jiménez and Campos-Ortega, 1979; Campos-Ortega and Jiménez, 1980). Also, White's mutant has a defect in ventral-nervous-system development that is strikingly similar to one of the lethal phenotypes caused by *l(1)X20*. Hence, this gene may have been reidentified. Some additional alleles of *vnd,* that now could readily be induced with chemical mutagens, may have the remarkable bistable effects on early neural and hypodermal development that were associated with the earlier mutation.

### Genetic Variants Affecting Postembryonic Development

*Visual System of Drosophila*

Genetic studies on the histogenesis of excitable cells in the fruit fly have frequently involved the visual system (reviewed by Campos-Ortega, 1980). The regularity of the patterning of cells in the eye itself and in the optic lobes has inspired many descriptive studies of the development and final morphology of this system (for example, Meinertzhagen, 1973, 1975). Thus, genetic manipulation of visual-system development can be considered against this extensive background of basic information. Each ommatidium or facet in the fly's retina contains eight photoreceptor cells, six outer cells ($R_{1-6}$) and two inner ones ($R_7$ and $R_8$) (see "Sensory Mechanisms"), and these cells are derived from the eye-antennal imaginal disk.

The possibility has been considered that there is a clonal relation among the cells of a given ommatidium; that is, during their histogenesis in the third larval instar, the 2,000 cells in the eye portion of the eye-antennal disk divide in a way that all photoreceptor cells in an ommatidium come from one "mother" cell. A more limited version of such a lineage hypothesis would be that all the outer cells or the two inner cells are within a given lineage. Descriptions of the developing retina in the third instar, when the cells are being produced and pattern formation is first evident, reveal a wave of mitosis and differentiation that moves posteriorly to anteriorly (Ready, Hanson, and Benzer, 1976). The mitotic wave is delineated by a furrow proceeding just in front of (anterior to) the dividing cells. An analysis of genetic mosaics in the developing eye (see figure 36A) showed that cell lineage plays little role in the histogenesis and pattern formation of the ommatidia (Ready, Hanson, and Benzer, 1976). The mosaics resulted from radiation-induced mitotic recombination (figure 36A) and marking of photoreceptor cells with the white ($w$) pigment mutation. In these eye mosaics, virtually any array of marked ($w$) and unmarked (wild-type red) cells was seen in the ommatidia—as opposed to what one would predict from the lineage hypothesis (that is, *all* cells within an ommatidia, or within the group of outer photoreceptor cells, would be either marked or unmarked). Thus, the cells appear to be recruited at random along the morphogenetic front whose existence was inferred from the description of the moving furrow. The position of the recruits is apparently what matters, not their ancestry. Another purely descriptive aspect of the early patterning of the retina is the appearance of a groove that bisects the developing eye disk into dorsal and ventral halves, anticipating the line of equatorial symmetry seen in the adult eye. But this symmetry is not generated by a clonal mechanism because the mosaics again showed marked cells in a single clone participating in the development of ommatidia above (dorsal to) and below the equator (Ready, Hanson, and Benzer, 1976).

It should be noted that the eye clones of Ready and coworkers (1976) were generated early in larval development. Whereas a cell that is marked at this early stage does not tend to be of the same genotype as any other specific cell in its (or any other) ommatidium, the developing eye cells, marked much later in larval development, seem as if they might show some clonal patterns (for example, Campos-Ortega and Hofbauer, 1977). This could mean that, when one of several presumptive photoreceptors, which have already been generated by early cell divisions, is marked, it tends to generate a specific member of the ultimate photoreceptor set during subsequent divisions that occur during the final stages of ommatidial histogenesis and cell differentiation. This hypothesis could be explained alternatively, however, if the groups of cells that showed coincident marking were only correlated by the fact that there had been more than one independent somatic crossover induced in the developing eye. Lawrence and Green (1979) addressed this matter in a study whose results strongly supported the "nonlineage" conclusion, as initially formulated (Ready, Hanson, and Benzer, 1976). In these experiments, Lawrence and Green induced clones of marked, white photoreceptor cells; but the somatic crossovers were deliberately designed to be rare because they were *within* the white gene complex on the X-chromosome (cf. Stern, 1969). It was still found that, in these single clones, the marked retinula cells are random combinations of the identifiably different photoreceptor types.

The controversy just discussed was not a mere "tempest in a teapot," in part because it points out that an interesting neural-developmental question can eventually be resolved by using increasing levels of genetic sophistication. Still, the actual mutants used in these clonal studies of receptor development are of relatively little intrinsic interest for neurobiology, except that the white mutation does cause defects in orientation and mating behavior (for example, Fingerman, 1952; Geer and Green, 1962). However, the induced mosaics plus the color marker used to detect and analyze them are very useful ways of studying developmental events retrospectively; that is, adults can be analyzed to deduce what occurred during the earlier developmental stages, during which determination of cell lineage versus positional mechanisms would be very difficult to achieve by direct examination of larvae and pupae at various stages. Genetic mosaics and markers have been used in studies of retinal development in a variety of organisms, including vertebrates such as mouse (Sanyal and Zeilmaker, 1977), axolotl (Van Deusen, 1973; Brun, 1978), and *Xenopus* (Conway, Feiock, and Hunt, 1980). Mosaicism induced by irradiation during the development of nematodes is now available for analysis of cell lineage, though it has not yet been applied to developing neural tissues (Siddiqui and Babu, 1980).

One gene whose mutations actually disrupt the histogenesis of the visual system (among other tissues) is *l(1)ddeg-3*. This X-chromosomal "disk-degenerate" mutation causes lethality (Stewart, Murphy, and Fristrom, 1972), due, at least in part, to the fact that all the major imaginal disks (examined in the third larval instar) are small and vestigial. Special attention has been paid to the eye-antennal disk and the presumptive optic lobes in this mutant (K. White, unpublished). For the latter, the anlagen of the lobes appear in the third instar not to have developed past their status in the first instar. The cell divisions in these anlagen that intervene between the early and late larval stages may not have taken place. Since the imaginal disks in *Drosophila* are connected to the CNS during the larval stages, one might imagine that a primary abnormality in the disks could induce this defect in the optic lobes (see "Inductive Tissue Interactions"). Indeed, in *l(1)ddeg-3* an abnormal eye-antennal disk can lead to abnormalities in apparently wild-type optic lobe anlagen, as revealed in genetic mosaics. However, the disk is not the only site of mutant action, as shown by the partial autonomy of *l(1)ddeg-3*'s action on the histogenesis of this part of the CNS. Again in genetic mosaics, defective optic lobe development has been seen in the presence of normal eye-antennal disk development. This mutation may therefore be deleterious to the formation of imaginal cells of the CNS as well as the imaginal disks.

There are other mutations that appear to affect the differentiation of particular excitable cells (photoreceptors) by preventing the formation of structures specific to these cells. The outer-rhabdomeres-absent (*ora*) and sevenless (*sev*) mutations result in defects in the peripheral photoreceptors and one of the central photoreceptors, respectively (Harris, Stark, and Walker, 1976; Campos-Ortega, 1980; see "Sensory Mechanisms"). For *ora,* the outer cells of an ommatidium are not totally defective, in that the cell bodies and axons are present; yet the light-sensitive rhabdomeres fail to form in each cell. The mutant is developmentally interesting because the defect does not appear to be a matter of postdevelopmental degeneration (unlike the receptor-degeneration mutant discussed earlier), and a specific cell type is affected. In *sev,* it is again only one cell type that is made to be defective; here, the central cell $R_7$ fails to form (Campos-Ortega, Jurgens, and Hofbauer, 1979). The defect in the central cell has been found to be due to an autonomous effect of the *sev* mutation. Thus, mosaic experiments (Harris, Stark, and Walker, 1976; Campos-Ortega, Jurgens, and Hofbauer, 1979) have shown that a genotypically *sev* cell never forms the normal retinal cell number 7; also, the genotype of neighboring cells within ommatidia that were themselves mosaic was uncorrelated with the presence or absence of cell number 7 (implying that its formation under the control of this gene does not involve particular cell interactions).

A gene affecting specific classes of cells in more central regions of the visual system is defined by the optomotor-blind variant (caused by a particular "break point" in a chromosome rearrangement). In this mutant, the behavioral defects discussed earlier are correlated with the absence of, or severe reduction in, specific giant fibers in one of the fly's optic lobes (Heisenberg, Wonneberger, and Wolf, 1978). Mutations causing more extensive and more central abnormalities in the fly's nervous system were induced by Heisenberg and Böhl (1979) and isolated on anatomical grounds. Several of the more interesting mutants, with consistent defects apparently occurring in only a portion of the brain, have characteristic patterns of abnormality in or near the dorsally located "mushroom-body" axons (Heisenberg, 1980; see Howse, 1975, for a general discussion of the mushroom bodies in insect brains). Other mutants, whose characterization is well under way, are caused by any of several mutations in the small optic lobe (*sol*) gene (Fischbach and Heisenberg, 1981; see "Behavioral Neurogenetics"). In the *sol^{KS8}* mutant, the three most central optic lobes are reduced to only about half their normal volumes. Many cell types appear to be normal in their intrinsic appearance in these mutant ganglia. Some cell classes, though, are missing. Still other cells (in the two most central lobes) send ectopic branches in a peripheral direction instead of making contacts only with their normal targets in the central brain per se. The aberrant, centrifugal branches in these particular cells in the mutant may be seeking connections that would be present (as in wild type) if the relatively peripheral lobe (the medulla) had its normal number of neurons.

These brain and optic lobe mutants further point out that differentiation of neurons can be selectively affected by particular genes. This is of intrinsic interest and also provides useful "genetic surgeries" for behavioral dissection.

*Development of Muscle Cells and Tissues*
Muscle cell histogenesis and differentiation have been found to be dramatically perturbable with mutants. Most of these mutants in *Drosophila* were discovered on the basis of abnormal positions of the wings (for example, Hotta and Benzer, 1972). Other cuticular abnormalities (for example, indented thorax) have proved to be good clues for defects present in the thoracic musculature. Histological study of these mutants (Hotta and Benzer, 1973; Deak, 1977) has shown that some of the sex-linked genes appear to be required for primary *formation* of muscles (flap wing, vertical wing); others are necessary for the primary *differentiation* of muscle cells (upheld, indented thorax); and one gene seems to be required for postdevelopmental *maintenance* of muscle structure once formed (heldup). Mosaic studies of these mutants have shown that most of them probably affect the muscles per se, since a fate-mapping analysis suggested that the defects are in embryonic cells (on the blastoderm) that are progenitors of muscles. Yet, since there was no direct marking of the genotype of muscle cells in these mosaics (Hotta and Benzer, 1972; Deak, 1977), neural or other defects that might perturb muscles by faulty cell interaction are not ruled out (see earlier discussion on fate mapping and also a later discussion on genetic muscular dystrophy). Indeed, one mutant (vertical wing) had its focus for muscle abnormalities mapped to a region of the blastoderm that very likely does not give rise to muscles (Deak, 1977).

Many of the *C. elegans* mutants isolated on the basis of uncoordinated movements (Brenner, 1974; Ward, 1977; Riddle, 1978) proved to have defective muscles. Subsequently, additional muscle mutants in this nematode were recognized by the criterion of defective muscle morphology (detected by both polarized light and electron-microscopic examination). Of the fourteen complementation groups that affect muscles (Waterston, Thomson, and Brenner, 1980; Greenwald and Horvitz, 1980) one is a gene coding for the myosin heavy chain (MacLeod, Waterston, and Brenner, 1977). The different alleles of the locus include null mutants, temperature-sensitive mutants, and one internal deletion within the gene. These genetic variants do not affect myosin in the pharyngeal muscles, nor do they affect one form of somatic myosin. Thus, the genetic results help to indicate that there are possibly three distinct forms of myosin in the nematode.

Among other muscle mutants in this organism are those in a gene coding for paramyosin (discovered as uncoordinated mutants; Waterston, Fishpool, and Brenner, 1977) and in several genes that affect the organization of myofibrils. Some of the latter mutations affect muscle development, shown in part by experiments using a temperature-sensitive mutation (Epstein and Thomson, 1974); others cause dystrophies of adult muscles (Brenner, 1974; Waterston, Thomson, and Brenner, 1980). These variants, then, are somewhat analogous to those in *Drosophila* genes that influence muscle formation or, on the other hand, maintenance of normal structure.

There are several recent and intriguing results on revertants of some of the muscle mutants, selected after mutagenesis of mutants with uncoordinated phenotypes. For example, some resulting second-site suppressor mutations partially restore the level of muscle proteins that is reduced or absent in the original uncoordinated mutant (Waterston and Brenner, 1978; Waterston, 1981). These suppressors are allele specific, not gene-locus specific, and suppress only null alleles. Such suppressor mutants behave genetically like classical "informational" suppressors in prokaryotes or yeast; that is, they may have altered transfer RNA molecules that can decode the molecular defect (during protein synthesis) in the messenger RNA from the mutated muscle gene. The molecular biology of such suppressor mutants may eventually prove useful in probes of the nature of mutational defects in a variety of developmental and behavioral mutants in the nematode (that is, in addition to muscle mutants). For instance, if a suppressor mutant is shown to allow the decoding of nonsense mutants in a gene known to specify a particular protein, the suppressor has been used to determine whether a mutant with no known protein abnormality, such as cell lineage mutation, is due to a nonsense, and thus totally null, mutation (see Horvitz and Sulston, 1980). Furthermore, suppression would indicate that the gene's primary function is to code for a protein product.

Other second-site suppressors in the worm, also recovered because of specific interactions with muscle mutants, are apparently not involved in protein synthesis in general. These suppressors (Riddle and Brenner, 1978) affect mutations in both the myosin and paramyosin genes. One of these second-site revertants is, in fact, a deletion of a gene or several very closely linked genes, setting this kind of suppression apart from informational suppression. These special suppressor factors may, therefore, identify additional genes that are specifically involved in the control of muscle assembly, function, or maintenance (see Riddle, 1980).

*Lineage Analysis of Excitable Cells in the Nematode*
The formation of the basic elements and ground plan of the nervous system is a problem that has been uniquely approachable in the post-embryonic development of the ventral nerve cord in the nematode. The small number of neurons (fifty-seven; White et al., 1976) and the visibility of their genesis and movement by interference microscopy (Sulston, 1976; Sulston and Horvitz, 1977) have permitted a description of nervous-system assembly at a level of detail unimaginable in *Drosophila* or the lower and higher vertebrates. Coupled with this high cytological resolution is an emerging analysis of the control by single genes of cell migration, proliferation, differentiative decisions of neuronal precursors, and autonomy of neuronal development (see review of Edgar, 1980).

At the time of hatching, the ventral nervous system (VNS) consists of fifteen neurons; later twelve neuroblasts or precursor (P) cells migrate from subventral positions into the ventral nerve cord, where they eventually give rise to most of the cells of the adult cord (Sulston, 1976; Sulston and Horvitz, 1977). Two genes have been identified that, when mutant, block the migration of these P cells into the VNS so that the adult nerve cord is not formed (Horvitz and Sulston, 1980; Sulston and Horvitz, 1981). The resulting incorrectly positioned cells generally do not divide, suggesting that either location in the ventral cord is necessary for P cell division or a single defect intrinsic to the P cells prevents both migration and division. Temperature-shift experiments using temperature-sensitive alleles of these P cell migration mutants have revealed that the temperature-sensitive period (cf. Hirsh and Vanderslice, 1976) coincides with the time of migration (E. J. Hess and H. R. Horvitz, unpublished).

The discovery of genes capable of exerting specific effects on neural migration is not in itself new; some of the best-characterized neurological mutations in the mouse are those that cause a failure of specific cell migrations. It is interesting that this kind of genetic step is shared by two animals as different in complexity as the nematode and the mouse.

Once the P cells have migrated into the ventral nerve cord, in normal development, they each undergo an identical set of divisions to produce six descendant cells: five neurons and one hypodermal cell (see figure 35). Not all of the descendants persist in every part of the ventral nervous system; some are programmed to die. Generally, however, lineally equivalent progeny differentiate into morphologically identical neurons. This pattern of histogenesis is invariant. Mutations that appear to block the different cell divisions have been identified. Preliminary studies of these mutants have provided indications of sequential decisions that a cell may go through on its way to becoming a specifically differentiated neuron (J. G. White, H. R. Horvitz, and J. E. Sulston, unpublished). When certain cell divisions are blocked, the undivided cell remaining at the mutant block is recognizable as one of the usual neuronal descendants. Thus, the specification of this remaining cell may indicate which neuron type a cell must become or pass through before it can go on to become another type. For example, if the division that produced the DAS and VD motoneurons is blocked, then the neuron that remains is of the VD type, suggesting that, in order to produce a DAS type, a cell must pass through a series of determinative stages, the last of which is being a VD type cell. These blocks are not local in their effects; the lineage is altered for the appropriate P cell descendants everywhere in the VNS.

The idea that a cell differentiates by means of a sequence of restrictive decisions has been advanced to explain findings from a number of developmental studies (Dienstman and Holtzer, 1975; Holliday and Pugh, 1975; Boyse and Cantor, 1978). A similar hypothesis involving binary choices has been formulated to explain a spectrum of mutations in *Drosophila* that specifically transform one type of cuticular tissue into another (reviewed by Kauffman, 1975; Ouwenweel, 1976; Kauffman, Shymko, and Trabert, 1978) and to account for an analogous set of transformations (called "transdetermination") of ectodermal percursor cells from *Drosophila* as a result of cell culturing (Kauffman, 1973, 1975). The results from the study of the nematode VNS provide support for these arguments by indicating that intrinsic, genetic mechanisms can control these apparent decisions at given steps, leaving behind one differentiated state and advancing to another. They extend the previous ideas by showing that, at least in the nematode, these controls extend down to the level of individual neurons.

Mutations that display considerable specificity in the alteration of cell division patterns affect the production of dopaminergic cells in the posterior lateral region of the animal. No division is blocked in this case; rather, it appears that one particular division repeats itself three times (Chalfie, Horvitz, and Sulston, 1981). Specifically, a daughter

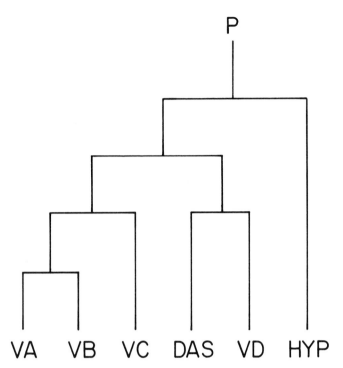

**Figure 35.**
Cell lineage of the ventral cord in the postembryonic development of the nematode. In
the development of *Caenorhabditis elegans,* twelve precursor cells (P) for cells in the
ventral nerve cord of the adult are present after hatching into the first larval stage. Each
cell migrates into the ventral nerve cord and undergoes a series of cell divisions, as
shown, eventually leading to the five neurons VA–VD, and a hypodermal cell HYP.
This pattern is repeated for different P cells (although there are some differences in cell
fates). [Horvitz, 1981]

cell seems to acquire properties normally associated with its own parent (as if a determinate switch has failed). One characteristic of the
parent cell is the generation of a single dopaminergic neuron; thus in
the mutants, multiple dopaminergic neurons are produced. The reiter-
ation of both lineage patterns and subsequent cell fates suggests a possible coupling between the differentiated phenotype of the cell and the
proliferative events that gave rise to it.

Lesions in the machinery of the cell cycle would be expected to disrupt development globally in all tissues. Two mutants of the nematode
appear to have general effects on the cell cycle, altering P cell divisions
as well as other cell divisions in the animal. In one of these mutants,
DNA replication fails to occur in any of the divisions of the P cells,
resulting in smaller and smaller descendant cells with less and less
DNA per cell, and a failure to differentiate (White, Albertson, and

Anness, 1978; Horvitz and Sulston, 1980; Sulston and Horvitz, 1981).
Another mutant has a reciprocal defect in that no postembryonic P
cell divisions occur at all, but DNA replication goes through six
rounds, as it would if divisions were proceeding unhindered (Albertson, Sulston, and White, 1978). Genetic dissociation of DNA replication from cell division, and the converse, have been observed
previously in cell cycle mutants of yeast (reviewed by Simchen, 1978).

Apart from their intrinsic interest as clues to the role genes play in
the developing nervous system, several of these mutations have also
been valuable as tools for asking about the consequences of cell deletions for development of the ventral nerve cord and other parts of the
nematode nervous system. The results give an overall picture of cellular autonomy in the development of these neurons; that is, elimination of individual cells (for example, through a block of one of the P
cell divisions) does not appear to alter the fate or connectivity of one
of its neighbors or targets (J. G. White, H. R. Horvitz, and J. E.
Sulston, unpublished). Ablations with laser microbeams of individual
cells have been used to ask the same question, with similar results (for
example, Sulston and Horvitz, 1977). In order to effect a deletion of
all of the neurons with postembryonic birthdays, the mutant that fails
to replicate its DNA was employed in a study of the neuron-target
interactions of a class of neurons that switches its connectivity post-
embryonically. These cells, which are present in the newly hatched
larva as neurons that innervate ventral muscles, reverse their polarity
and become neurons that innervate dorsal muscles during early post-
embryonic development. It was thought that this "rewiring" might
arise in response to the postembryonic cells, but the cells underwent
their normal rewiring even in the DNA replication mutant that failed
to produce mature target cells for the neurons (White, Albertson, and
Anness, 1978).

## INDUCTIVE TISSUE INTERACTIONS

Interactions between excitable cells are not only crucial for their functioning in adult stages but also for their earlier development. This
seems obvious, in that nerve cells make connections with other nerve
cells, muscle cells, or endocrine cells during development. More sub-
tly, it seems also to be true that the initial stages of the formation of
such connections may have an important influence in furthering the
cell differentiation of the target cells and the overall pattern formation
of all the cells involved. Thus, the formation of certain connections
may do more than simply *become* a pattern; it may also specify the
development of additional patterns among other cells. Mutants have
proved to be very useful in demonstrating cell interactions in the

developing nervous system. Without mutants, one might merely observe a developing axon contact a presumptive target cell, and imagine that later changes in the target were induced by the contact from the other cell (for example, Levinthal, 1974). Actually, however, the observed contacts may not be causal at all, or influences may be flowing in a direction opposite to the one imagined.

## Interactions during Visual System Development of Invertebrates

Examples of how the dilemma discussed in the above section was attacked in arthropods have been reviewed by Palka (1979b), Anderson and coworkers (1980), and Macagno (1980). An especially cogent example comes from work on the *Drosophila* visual system by Meyerowitz and Kankel (1978). They learned that certain mutations that disrupt the differentiation and patterning of cells in the retina also grossly disrupt the patterned wiring of neurons in the optic lobes. The authors realized, though, that in these mutants—rough eye (*ro*), glass eye (*gl*), and Glued eye (*Gl*)—the optic lobes as well as the cells in the retina are mutant. The wiring in the target cells could be due to inductive influences or intrinsic genetic defects. The former result was found. In genetic mosaics (figure 36A), the optic lobe phenotypes of all three of these mutants were correlated with the retinal genotype and phenotype (figure 36B). For instance, a patch of mutant *ro* or *gl* tissue in the retina would be associated with disordered optic lobe wiring beneath it. In the dominant *Gl* mutant, a wild-type patch of retina specifies a normal wiring in the genotypically mutant neurons beneath it (figure 36C). Thus, some information on wiring must pass from the retina to the optic lobes.

These effects on the organization of visual pathways, due to defects in the eye, are of the same general type as seen in certain vertebrate mutants, such as those with reduced eye pigment (discussed under "Nerve Specificity"). Thus, more central defects are assumed to "cascade" from a specific alteration of the periphery. Mosaics have not been extensively used in these systems, however, to determine whether genetically normal central cells are truly induced to miswire under the influence of mutant cells in the eye, though there are other indications that this is the case.

The mechanisms involved in the cell interactions demonstrated with the fly eye mutants are unknown; for instance, is there faulty pattern formation or faulty maintenance of pattern once formed? The mutants or mosaics have not been examined during developmental stages, let alone studied biochemically. In this light, the actual products from these genes may prove to be interesting. The reason is that *ro, gl,* and *Gl* may be, at least in part, *specifically* concerned with eye-

PARENTAL CELL
GENOTYPE

DAUGHTER CELL
GENOTYPES

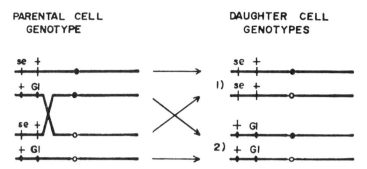

**Figure 36A.**
Induced somatic crossing over in developing *D. melanogaster*. The third chromosomes
are shown heterozygous for the dominant mutation Glued (*Gl*) that perturbs the pattern
of facets in the eye and the patterning of neurons in the optic lobes. These chromosomes
are also heterozygous for the recessive eye-color marker sepia (*se*). At a known time
after egg laying, the animals are exposed to about 1000 rad of ionizing irradiation,
which will induce a mitotic crossover, as shown, in a few percent of the treated individuals. If the crossover event occurs between the location of the gene under study (*Gl*)
and the centromere (closed and open circles), then normal mitotic segregation of centromeres can lead to homozygosis for the respective alleles of the heterozygous gene
loci; that is one daughter cell can be $Gl^+/Gl^+$ (and *se/se*), and the other *Gl/Gl* (and
$se^+/se^+$). [Meyerowitz and Kankel, 1978]

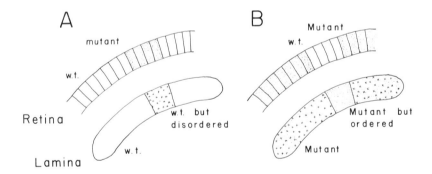

**Figure 36B.**
Summary of the results of Meyerowitz and Kankel's (1978) mosaic study of eye-brain
mutants in *Drosophila*. Wild-type tissue is shown unshaded and mutant is stippled;
disordered lamina is indicated by crosses. (*A*) Wild-type (w.t.) but disordered lamina is
present under a patch of mutant retina produced by mitotic recombination. (*B*) The
lamina is all mutant, but the portion of it under the w.t. retinal patch is rescued from
disorder. [Palka, 1979a]

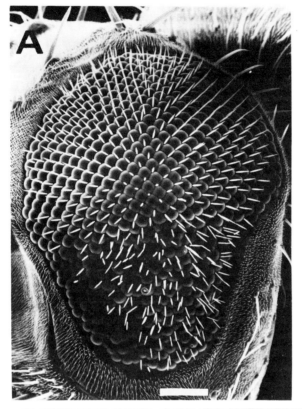

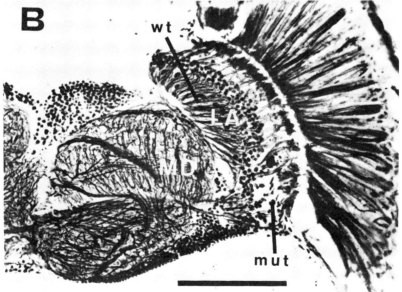

brain interactions; this is because only for a small fraction of the many mutants does a disrupted array of facets in the eye lead to wiring abnormalities in the brain (Meyerowitz and Kankel, 1978).

It should be noted that another mutant with aberrant gross morphology and wiring in *Drosophila* optic lobes—a small optic lobe mutant of Fischbach and Heisenberg (1981)—has been analyzed in mosaics. In this mutant, unlike what is found in the three other mutants just discussed, the small and miswired optic lobes in *sol*$^{KS58}$ are not the result of any effects of this mutation in the eye.

The lethal mutant *l(1)ddeg-3* has been used to demonstrate somewhat broader types of cellular interactions in the visual system, concerned more with overall tissue formation than with specific pattern. While the mutation apparently affects both imaginal disk and brain development in an autonomous fashion, the effects on the disks—at least the eye-antennal disk—can induce defects in parts of the brain. Thus, in genetic mosaics, a mutant eye disk is always associated with an abnormal optic lobe whether the lobe itself is mutant or normal in genotype (K. White, unpublished). The eye and the optic lobes are not very closely related developmentally (Kankel and Hall, 1976); hence, several of the cases of abnormal optic lobes seen in these mosaics must have been induced by abnormal input from the mutant periphery, even though the phenotypically aberrant parts of the brain were genotypically normal. This induced defect is different from the abnormality of optic lobes seen in uniformly mutant larvae. In the former case, development of the presumptive optic lobes appears normal until the third larval instar. Therefore, the induced defect is different than that which occurs when both the inductive and intrinsic controls of optic lobe development are defective.

**Figure 36C.**
Mosaic expression of the Glued mutation in *D. melanogaster*. Mosaics for this third chromosomal mutation were generated by irradiation-induced somatic crossing over. (*A*) Mosaicism in the eye, observed by scanning electron microscopy. The top part of the figure (dorsal) is an induced clone of $Gl^+/Gl^+$ tissue, while the bottom part (ventral) is still heterozygous for this dominant mutation ($Gl/+$). The scale bar represents 60 $\mu$m. (*B*) Mosaicism in the optic lobes, observed after freeze-sectioning in the horizontal plane and silver staining. The non-Glued clone in the eye of this mosaic (a different fly from that shown in A) is toward the top right-hand portion of the bright-field micrograph (that is, occupying most of the anterior portion of the eye); the mutant $Gl/+$ clone is at the bottom right (posterior). The $Gl^+/Gl^+$ patch in the eye is associated with phenotypically wild-type (wt) tissue in the optic lobe (the lamina, LA), which is just proximal to the eye. This lamina projects through an optic chiasma to a medulla (MD), which is intermediate in phenotype between Glued and wild type. The more posterior part of the lamina, associated with the Glued tissue in the eye, is mutant (mut), that is, Glued phenotype. The scale bar represents 50 $\mu$m. [Meyerowitz and Kankel, 1978]

An internal marking technique (Kankel and Hall, 1976) could determine whether a defect solely in the CNS in *1(1)ddeg-3* mosaics could induce a defect in the imaginal disks. However, this kind of inductive interaction (in the opposite direction from the flow of information in the physiological functioning of the adult) is usually thought *not* to be necessary in insect development. This is because the eye disk appears to differentiate normally when entirely separated from its usual location near its targets in the CNS (Chevais, 1937; Eichenbaum and Goldsmith, 1968).

Cell interactions of sensory and central cells involved in the development or maintenance of the normal form of these tissues are very powerfully demonstrable using mutants and mosaics. Other kinds of intervention are more commonly used to infer particular interactions of excitable cells, such as surgery of axons away from their target cells and tissues (reviewed by Jacobson, 1978; Anderson, Edwards, and Palka, 1980); reciprocal transplantation of neural and mesodermal tissue involved in development of limbs in the amphibians (Harrison, 1934); removal of the developing antenna in the moth *Manduca,* thus eliminating its input to the antennal glomerulus in the brain (Hildebrand, Hall, and Osmond, 1979); and ultraviolet microbeam surgery of specific photoreceptor cells that are in the process of developing connections with the presumptive brain in *Daphnia* (Macagno, 1978, 1980).

**Interactions during Visual System Development of Vertebrates**

Many investigations of tissue interactions have been carried out on the visual system of amphibians, owing to the availability of mutants and variants. The relative ease of performing experimental embryological studies on these animals and the kinds of information that they have yielded about the developmental dysfunctions in these mutants are well exemplified by the work on the eyeless mutant of axolotl. The most outstanding feature of the eyeless (*e*) mutant, which was discovered as a spontaneously occurring recessive mutant (Humphrey, 1969), is the failure of optic vesicles to form (Humphrey, 1969; Van Duesen, 1973). These mutants are also darkly pigmented and sterile. Van Deusen (1973) undertook a series of experiments involving transplantation of embryonic tissue between eyeless mutants and normal axolotls in order to identify which component of the inductive system for optic vesicle formation is directly affected by the mutation. His most instructive experiment is outlined in figure 37. Presumptive ectoderm or presumptive chordomesoderm was grafted reciprocally between eyeless and normal early gastulae. The results of these experiments placed the fault with the ectoderm and not the inducing mesoderm, since normal mesoderm could not induce eye formation in eyeless ectoderm and eyeless mesoderm was capable of inducing normal ectoderm. A further refinement was to determine whether the neural ectoderm was not being induced or the overlying epidermis (transplanted simultaneously with the neural ectoderm) was somehow suppressing optic vesicle formation. Brun (1978) showed the latter alternative to be the case by transplanting presumptive normal or mutant lens tissue over presumptive eye neural ectoderm at the neural fold stage of development. The result is that the fault seemed to lie in the epidermis. Mutant epidermis will suppress eye formation when grafted over normal diencephalon, and normal epidermis will allow eye development of mutant ectoderm (Brun, 1978). This result may make some sense in light of Ulshafer and Hibbard's (1979) electron-microscopic finding that the space between the diencephalon wall and head ectoderm in eyeless embryos is filled with mesenchyme. It is possible, for example, that the epidermal cells are responsible for clearing away the mesenchymal cells over the incipient optic vesicle and that the absence of overlying mesenchyme is necessary for the evagination of the vesicle.

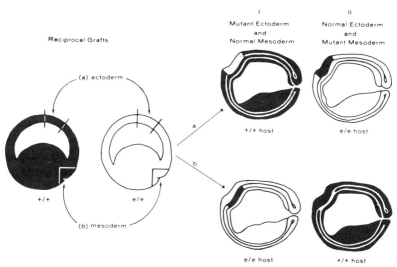

**Figure 37.**
Experimental scheme for determining which component(s) of the eye-forming complex in the axolotl is primarily affected by the eyeless gene. Reciprocal orthotopic exchanges of either the presumptive anterior neural plate ectoderm (a) or presumptive mesoderm (b) between white eyeless (*d/d; e/e*) and dark normal ( + / + ; + / + ) early gastrula produce chimeras in which normal mesoderm is tested for its ability to induce mutant ectoderm (Ia, Ib), and mutant mesoderm is tested for its ability to induce normal ectoderm (IIa, IIb) to differentiate eyes. [Van Deusen, 1973]

The amphibian nervous sytem has also been the subject of a kind of genetic analysis of inductive and interactive mechanisms that does not involve mutants but, rather, the genetic differences between related species. These studies are done in mosaics, in which part of the animal is of one species and the rest of another. (Amphibian mosaic studies with particular relevance to the nervous system have been recently reviewed by Harris, 1979.) With such mosaics one can ask, What are the potencies of the various embryonic germ layers, and which parts of the organism are able to induce and organize the nearby tissues of another species? One can also learn what developmental chemistry may be common to different species. Chimeric experiments have, for example, told us a great deal about the development of the eye.

The original size of the optic vesicle is a property of the inducing mesoderm and not an intrinsic property of the ectoderm that will, in fact, form the eventual eye cup. Rotmann (1939) and Balinsky (1958) transplanted gastrula ectoderm into the future eye region of the neural plate between different species of anurans and urodeles. The eye rudiment that was then induced from the graft was generally of the same size as the normal host eye rudiment. In other words, the inducing mesoderm decided how many cells or how much territory of the neural plate should be set aside for making an eye.

Later in development, however, the foreign optic vesicle, although initially of the size prescribed by the host, begins to grow disproportionately for the host at a rate of growth directed by the genetic instructions intrinsic to the eye tissue (Twitty, 1940, 1955). Eyes, like limbs, grafted between two species of urodeles essentially exhibit the growth rate of the donor species (Harrison, 1924, 1929; Stone, 1930; Twitty and Schwind, 1931). The size increase of the total organ appears to be relatively unaffected by the growth of its immediate environment. Thus, a *tigrinum* eye becomes disproportionately large for its *punctatum* host, and a *punctatum* eye grafted to a *tigrinum* preserves its small size, in spite of the fact that it is nourished by a much larger and more rapidly growing host (figure 38; Harrison, 1929; Stone, 1930; Twitty, 1930).

Another important factor for retinal development is the lens. When a potential size maladjustment is made by combining the lens from one species with the optic vesicle of another, the two tissues exert a mutual influence on each other and an intermediate size is reached (Harrison, 1929; Ballard, 1939).

The size and rate of growth of the eye are thus directed by specific intrinsic and extrinsic factors. The eye, in turn, influences the surrounding tissues. The extrinsic ocular muscles and skeletal structures adjust their size to that of the eye (Twitty, 1932). When a *tigrinum* eye is grafted to a *punctatum,* the host eye muscles grow beyond their nor-

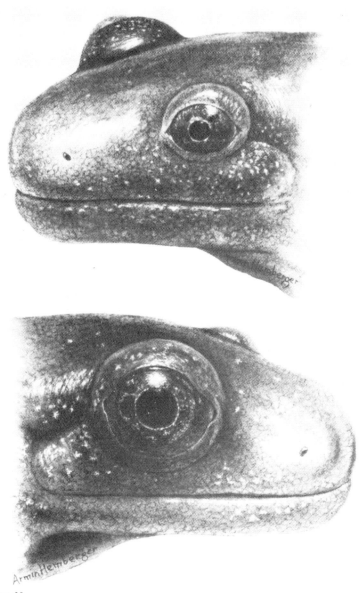

**Figure 38.**
A small-eyed adult *Ambystoma punctatum* bearing a transplanted growing right eye (which is larger) from a *tigrinum*. Transplant performed in the early larval stage of development. [Stone, 1930]

mal size, by an increase in the number of fibers, in order to keep a reasonable proportion with the giant eye to which they are now attached (Twitty 1932, 1955). Furthermore, heteroplastic giant (or miniature) eyes with greater (or fewer) than the normal number of retinal ganglion cells lead to cellular hyperplasia (or hypoplasia) in the contralateral midbrain. It is not known whether the midbrain hyperplasia represents induced mitoses or absence of cell death, nor whether there are any trophic substances working in this system.

There are many examples from mammals on the use of mosaics to sort out inductive or interactive influences in neural development. Mosaic mammals, called allophenics, chimeras, or tetraparentals, are generated by mixing cells from two separate embryos, followed by reimplantation into a pseudopregnant mother, after which the development of the now mosaic embryo resumes. The details of the procedure, pioneered by A. K. Tarkowski and by Mintz, and its many applications (which extend far beyond neurobiological usage) have been reviewed several times (Mintz, 1974; Mullen, 1977a; Gardner, 1978; McLaren, 1978). Since one mixes cells from two early embryos (usually at the eight-cell stage, each embryo resulting from a separate and genetically defined cross), the mosaic formed by aggregating the two kinds of cells develops with mutant and normal cells simultaneously present in a single individual.

Complex genetic techniques are not involved in the construction of these mosaic chimeras (compare the X-chromosome-loss mosaics in *Drosophila* or the X-chromosome-inactivation mosaics used by Guillery and coworkers, 1973, to analyze nerve miswiring induced by an albino mutation, to be discussed). This is advantageous because one can easily introduce an external morphological or internal cell-marker variant into a strain carrying a neurological mutation.

Some relatively early examples of the use of these chimeric techniques involved studies of mammalian eyes. An effect of mutant pigment cells in the retina of the rat, expressing a retinal degeneration mutation, was revealed with this type of mosaic. In these experiments (Mullen and LaVail, 1976), the mutant genotype of the pigmented epithelial cells was found to be necessary and sufficient for inducing degeneration of photoreceptor cells. Chimeras were constructed using *rdy* cells, which also carried a nonpigmented marker for retinal ephithelial cells, and *rdy*+ pigmented cells. Patches of abnormal and degenerated photoreceptors were present only when opposite the marked *rdy* cells of the pigment epithelium. There was no marker for the genotype of the photoreceptor cells, and thus it seemed possible that both the epithelial and photoreceptor cells might have to be mutant in order that degeneration occur. Yet *all* the patches of mutant pigment epithelium (out of more than 200 examined) had degenerating and abnormal photoreceptor cells opposite them. A reasonable fraction of the marked patches would, no doubt, have been associated with $rdy^+$ photoreceptors, leading to the conclusion that an induced defect in the light-sensitive cells is the cause of their degeneration. Hence, the overt abnormality in the photoreceptor cells is independent of their genotype.

The $rdy$ epithelial cells were known to express a defect related to turnover of photoreceptor cells: The outer segments from the latter cell type, when shed, are phagocytosed by the epithelial cells, and this process is defective in the mutant. The mosaic study showed that the defect is intrinsic to the epithelial cells. It appears to be rather specific because the $rdy$ pigment epithelial cells can phagocytose gratuitous materials, such as carbon or polystyrene beads.

Genetic retinal degeneration ($rd$) in the mouse has also been studied, again using chimeras (see review by Mullen, 1978). One feature of the patchy degeneration is that there are regions of intermediate retina where the degeneration is incomplete and the retina is abnormal. Unlike the case of $rdy$ in rat, the mosaics studied with respect to $rd$ in the mouse show no correlation between retinal degeneration and the genotype of the pigment epithelium. A general feature of mouse retinal mosaics, with respect to mutant patches that are merely marked or that also express a degeneration mutant, is that there is the distinct suggestion of radiating patches in the eye. This is evidence for clonal development, in which most of the descendants of the marked cell remain fairly contiguous, as if they had been in some way programmed to form a sector of the retina. But, whereas the distribution of marked and unmarked cells is not random (Sanyal and Zeilmaker, 1977), the clones are not so distinct and delineated as to rule out extensive cell mixing. Nor does the nonrandomness necessarily imply anything about the program of development in the retina other than the passive displacement of cells during proliferation.

It should be noted parenthetically that hereditary retinal degeneration has also been studied in invertebrate eye mosaics. An X-linked photoreceptor-degeneration mutant in *Drosophila* (see "Sensory Mechanisms") was shown by Harris and Stark (1977) to have an intrinsic defect in the outer retinula cells of each ommatidium, in that a mutant genotype of nearby pigment cells could not induce wild-type photoreceptor cells to degenerate.

### Interactions during Development of the Central Nervous System

Several mutants of the CNS in mouse have been discovered by observing abnormal behavior of neonates. The mutants show aberrant postures and movements and are loosely named after their overt defects:

hence, such mutant designations as staggerer, reeler, and weaver (Sidman, Green, and Appel, 1965; Sidman, Cowen, and Eicher, 1979). Many of the single-gene mutations responsible for these abnormalities have been shown to affect at least the postnatal development of the brain. Very early histogenesis, cell differentiation, and pattern formation in prenatal development are not dramatically perturbed in most of the mutants, which might be expected because the mutations do not cause lethality (and they essentially could not, given the way they were detected).

The different mutants affect the development of the brain in different ways, and the studies, aimed at determining the site or sites of mutant gene action—and, by inference, wild-type gene action—have provided useful information on cell lineages, movements, and interactions during terminal stages of neural development. The rather straightforward descriptions of the cerebella to be discussed here (focusing on only a limited number of the extant mutants) have revealed that (1) Purkinje cells degenerate in the Purkinje-cell-degeneration (*pcd*) mutant, all having done so by about 1 month after birth; (2) granule cells degenerate and Purkinje cells lack spines in the staggerer (*sg*) mutant; (3) the cerebellum is reduced in size, lacks folia, and has malpositioned cells in the reeler (*rl*) mutant.

These overt defects need not be the primary sites or foci of gene action, because there are, of course, many cell interactions occurring among these cerebellar neurons, in conjunction with the considerable degree of migration of various cell types occurring postnatally. Certain cell contacts have been observed, for instance, between migrating granule cells and glial cells, and these could be important in controlling migration. Hence, nonmigrating cells in a mutant might *not* be intrinsically defective. Abnormal migration can, of course, lead to abnormal positioning of the various layers of cell types. In addition to the possibility of abnormal migrating or guiding cells causing the malpositioning, it is also possible that a general "field" property of the mutant cerebellum affects many features of these cell movements simultaneously. Once migration is complete, further cell interactions remain important, as a particular class of cell establishes contacts and synapses with its target cells of another particular class. The failure to establish normal contact could be due to aberrant input cells, target cells, or both. If the target cell is defective, then the absence of, or improper contact with, the input cell could eventually cause that cell to degenerate. Thus, normal cell contact has to be initiated properly. It also needs to be maintained as part of a continuing communication between input and target, mediated by basic synaptic transmission or also, perhaps, some kind of hypothetical trophic interaction.

By constructing chimeric mice, the *pcd* mutant has been found to

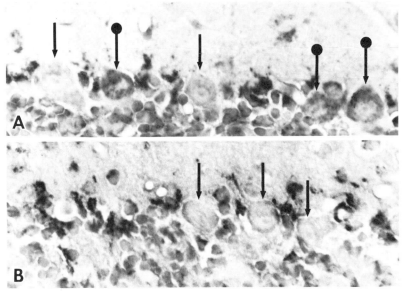

**Figure 39.**
Sagittal sections (7 μm) of cerebellar cortex of mouse stained histochemically for
β -glucuronidase and counterstained with methyl green. The mice were perfused with
cold (0–4 °C) 4% formaldehyde in phosphate buffer. The brains were processed and
stained as described by Feder (1976). The sections were stained for 3 days at 37 °C with
fresh substrate added daily. In these photomicrographs, the red precipitate at the sites
of glucuronidase activity seems dark gray. (*A*) Cerebellar cortex of a $Gus^b/Gus^b$ ←→
$Gus^h/Gus^h$ chimaera. Some of the large Purkinje cells are stained (arrows with dots),
indicating that they are from the $Gus^b$ component; others are not stained (arrows) and
are, therefore, $Gus^h$. (*B*) Cortex of a $pcd/pcd$ $Gus^b/Gus^b$ ←→ $+/+$ $Gus^h/Gus^h$
chimaera. None of the surviving Purkinje cells is stained (arrows), indicating that they
are from the $+/+$ $Gus^h/Gus^h$ component. To the left is a gap left by degenerated
Purkinje cells. In both *A* and *B* the other intense staining in the Purkinje cell layer is
probably in Golgi ephithelial cells (Bergmann glia). [Mullen, 1977]

exhibit degeneration of the Purkinje cells, resulting from *intrinsic*
defects in these cells as opposed to abnormal outside influences
(Mullen, 1977b). Some of the embryos resulting from mixing *pcd* and
wild-type cells developed into neonates that showed at least a partially
mutant phenotype in the cerebellum (Mullen, 1977b; figure 39).
Whether a given cell in the "mixed" brain expressed the neurological
mutant was determined through the use of the histochemical marker
β-glucuronidase, the structural gene for which (*Gus*) has high-activity
and low-activity alleles. The mutant or wild-type Purkinje cells could
thus be "tagged"; it turned out that the only surviving Purkinje cells
were marked as wild type, strongly suggesting that the mutation acts
autonomously in these cells. That is, whereas it is not clear whether *all*
the wild-type cells survived, apparently none of the mutant cells could

be "saved" by any kind of input from normal cells or substances in the mixed cerebella.

The patterns of mosaicism in the *pcd* chimeras—in addition to the site of gene action—reveal an important and well-known feature of brain development in mammals. The mosaicism is "finegrained," without any contiguous clones of many marked or unmarked cells. (This is, of course, also observable in mosaics not involving a neurologically defective mutant.) Clearly, then, there is much cell movement and mixing during brain development; hence; mutant-normal dividing lines are essentially never seen (unlike the case of mosaic brains in *Drosophila*). Thus, in chimeric *pcd* brains, the distribution of surviving, wild-type Purkinje cells was essentially random; there were no regions where their density was normal, nor were there any places in which they had all degenerated (Mullen, 1977b).

Mosaics constructed with respect to the staggerer mutant have suggested the importance of an interaction between the granule cells and their Purkinje cell targets (Herrup and Mullen, 1979). In the mosaic cerebella, the Purkinje cells that were normal in size and position were wild type; those that were small and out of position had the *sg* genotype. Thus, the mutation affects the target Purkinje cells, and the degeneration of the contacting granule cells could essentially be induced by the defects in their targets. However, it is also possible the *sg* affects the granule cells directly, as well as the Purkinje cells. This question cannot be resolved until a cell marker for granule cells is established (the enzyme controlled by *Gus* is not uniformly expressed in these cells.) In addition to the possibility of the faulty cell interaction in *sg* (that is, eventual degeneration of granule cells) being triggered by an initial defect in the Purkinje cells, there also may be a defective interaction per se in the initial aspects of the input-target contacts. Such a suggestion comes from the finding that, in general, cerebellar cells from *sg* individuals exhibit persistence of a characteristic associated with embryonic cell surfaces (Trenkner and Hatten, 1979). This was revealed by the demonstration that, contrary to wild-type cerebellar cells, dissociated *sg* cells do not show a loss during embryogenesis of agglutinability by plant lectins (see also review by Mallet, 1980).

The reeler mutant shows a host of developmental abnormalities at various stages and in many parts of the brain. Chimeras involving *rl* showed that a large number of Purkinje cells can be malpositioned (as most of them are in the fully mutant animal; figure 40). The important finding from the mosaics was that not all of the malpositioned cells were genetically *rl,* and some *rl* cells were correctly positioned (Mullen and Herrup, 1979). The marked cells in these mosaic cerebella did not apparently show an "intermediate" phenotype, in that the wild-type *rl*

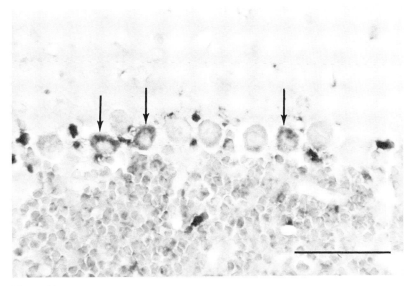

**Figure 40.**
Section of mouse cerebellar cortex from a reeler chimera ($Gus^b/Gus^b$ $rl/rl$ ◄─►
$Gus^h/Gus^h$ $+/+$). This area of cortex appears normal with molecular layer at the top, single layer of Purkinje cells in the middle, and granule cell layer at the bottom. The section was stained for $\beta$-glucuronidase and counterstained with methyl green. The red precipitate at sites of glucuronidase activity appears black in these photographs. Three Purkinje cells are stained (arrows) and are, therefore, $Gus^b/Gus^b$ $rl/rl$ in genotype, yet normal in phenotype (i.e., position). Conversely, there were $Gus^h/Gus^h$ $+/+$ cells exhibiting the reeler phenotype (i.e., aberrantly positioned) elsewhere in the same cerebellum. Bar represents 50 $\mu$m. [Mullen and Herrup, 1979]

Purkinje cells were either in their correct positions or else in the abnormal location characteristic of such cells in the full mutant. A tentative and admittedly general conclusion, then, is that this mutation affects pattern formation in the brain by altering some general positional signals that can affect several cell types by extrinsic influences. At the least, the defective or normal positioning of Purkinje cells is not affected by intrinsic cellular genotype.

Some of the classic mosaic experiments on the development of central tissues come from early work on amphibians. Amphibian species may differ from each other in pigmentation. Such a difference can be used as a reliable and long-lasting marker between transplant and host tissue, and this makes it possible to decide the species origin of a given cell in a chimeric individual. The primary inductive origin of nervous tissue was uncovered by such a mosaic analysis (Spemann and Mangold, 1924). Before these experiments, it was known that the dorsal lip of the blastopore was critical in determining the major axes and development of the embryo. A dorsal lip transplanted ectopically on the embryo, unlike other ectodermal transplants that join in the

development of the new surroundings, results in the formation of a
small secondary embryo with a neural tube, notochord, and somites
(Lewis, 1907; Spemann, 1918). At first the significance of these
experiments was unknown, since it was assumed that the secondary
embryo was formed from the material of the implant. That this was
not the case was first shown by Spemann and Mangold (1924), who
transplanted the dorsal lip heteroplastically between the gastrulae of
differently pigmented *Triton* species. The results of this experiment
led to the development of the concept of a primary organizer in neural
induction. The dorsal lip invaginates and becomes mesodermal or
chordomesodermal, and from its position inside the gastrula it induces
the overlying ectoderm of the host animal to form the spinal cord and
brain (figure 41).

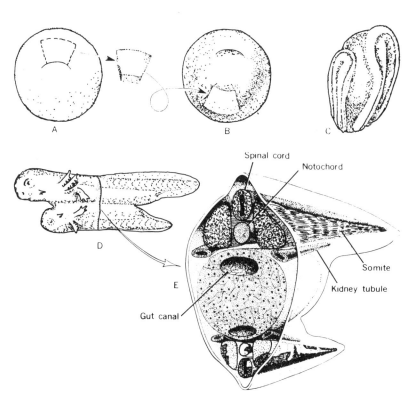

**Figure 41.**
Induction of secondary embryo in the darkly pigmented *Triton taeniatus* by transplantation of dorsal lip of the blastopore from the light-colored *Triton cristatus*. This
experiment, first performed by Spemann and Mangold (1924), shows that the graft
organizes the tissues of the host to participate in the development of the secondary
embryo. (Note the dark tissue in the spinal cord.) [Twitty, 1966]

Spastic (*sp*) is a neurological mutant in the axolotl with some similarities to the reeler mutant of mouse. The *sp* mutation leads to coiling and thrashing motor patterns, an inability to swim with the usual sinusoidal oscillations of these amphibia, and equilibrium dysfunction (reviewed by Ide, 1978). Several different approaches have been used in attempts to determine the focus of the defect. First, the onset of defective swimming in the mutant is at a stage corresponding to the onset of function in the midbrain and cerebellum (see Ide et al., 1977a, and references to previous work therein). Second, phenocopies of spastic can be produced by lesioning the vestibular root or the vestibulocerebellum in normal animals (Ide et al., 1977a). Third, the location of Purkinje cells in the mutant vestibulocerebellum is shifted ventrally (Ide, Tompkins, and Miszkowski, 1977b). Fourth, electrophysiological studies have shown that a certain class of vestibulocerebellar units responding to tilt is present but displaced from its normal position (Ide, 1977).

Chimeric studies on the spastic mutant have preliminarily indicated that the focus is in the cerebellum. In these experiments (C. F. Ide and R. K. Hunt, unpublished), unilateral grafting of presumptive cerebellum from normal axolotls into the neural plate of spastics (at the neurula stage) seems to correct the equilibrium dysfunction on the side ipsilateral to the graft. Bilateral cerebellar transplantation corrects both the equilibrium and the swimming disabilities. Unfortunately, in these mosaic experiments the transplantations of normal tissues into the developing spastic individuals did not rescue the mutants with respect to the very low viability caused by *sp;* nor did any of the behaviorally rescued spastics reach adulthood. Thus, it may be that there are other, as yet undetected, foci influenced by this genetic variant.

## Interactions Involving Development and Maintenance of Muscles

A very specific inductive interaction during development of muscle tissue in amphibians was dramatically reinforced by the results of mosaic experiments on a mutant in the Mexican axolotl. Here, the cardiac mutation results in a nonfunctional heart, although the initial development of this organ is ostensibly normal (see review by Malacinski and Brothers, 1974). The mutational defect appears to be quite specific since skeletal muscles are unaffected. The heart muscles, though, show a lack of organized sarcomeric myofibrils. Mosaic work by Humphrey (1972), discoverer of the mutant, indicated that the overtly defective tissue is not the site of the primary genetic defect. Mutant heart primordia, transplanted into the heart regions of normal recipients, function normally; yet wild-type presumptive hearts are induced to develop defectively when put into mutant hosts. This suggests that anterior endodermal tissue—known to be involved in heart development in a variety of vertebrates (see references in Lemanski, Paulson, and Hill, 1979)—fails to induce cardiac muscle differentiation in the mutant. Lemanski and coworkers (1979) have strongly reinforced this suggestion through a mosaic experiment performed in vitro. They took anterior endodermal tissues from developing wild-type salamanders and cultured them with presumptive hearts from mutant animals. The anterior endoderm (but not other tissues, such as developing somites or posterior endoderm) caused about 80% of the mutant hearts to develop normal beating (culturing mutant heart tissue alone led to no normal contractile behavior). Moreover, the mutant heart tissue was induced to develop normal morphology as viewed ultrastructurally. Once more, then, the mosaic approach has provided a considerable degree of understanding of the nature of a mutational defect.

Suggestions have been made concerning more specific interactions involving muscles, that is, the influence of motoneurons on the development and maintenance of skeletal muscle integrity. Many of these hypotheses have been developed through the use of muscle mutants (Sidman, Cowen, and Eicher, 1979) and genetic mosaics in these vertebrates. Often, however, the experimental interventions made to generate the mosaics have been performed relatively late in the development of muscle tissues, and the results have led to some difficulties in interpretation of possible cell interactions. For instance, in neuromuscular work on the mouse, transplant experiments with tissues that are putatively involved in *genetic muscular dystrophy,* and that have already developed to relatively late stages, have sometimes implied an absence of influence of mutant nerves on wild-type muscles (Hamburgh et al., 1975; Law, 1977). To obtain an unambiguous answer, mutant and normal tissue must be juxtaposed from very early developmental stages onward. Thus, an early inductive effect, or one that requires "chronic" defects of induction, can be shown. This kind of experimental situation has been achieved in vertebrates by using genetic mosaics. Peterson (1979) and Peterson collaborators (1979) have made mouse chimeras, starting at very early stages, between dystrophic and normal strains and have concluded that there is an extrinsic influence on the development of muscular dystrophy. Indeed, these dystrophic mice have been shown to have morphological abnormalities in certain regions of their spinal roots and cranial nerves (Bradley and Jenkinson, 1973; Stirling, 1975), as well as, of course, in their skeletal musculature. The necessity of functional neuromuscular connections for maintenance of the integrity of adult muscles is well known (Jacobson, 1978). Therefore, perhaps the nerves are the primary site of the genetic defect and induce the muscular abnormalities.

Peterson and colleagues have made use of muscle enzyme markers involving expression of the malic enzyme (*Mod-1*) or glucose phosphate isomerase (*Gpi-1*) in their studies of the site of action of the dystrophic (*dy*) mutants in the mouse. Different electrophoretically detectable alleles of the *Mod-1* or *Gpi-1* genes were used; thus a rather definitive identification of tissue genotype was possible.

An analysis of muscle genotype and phenotype in the adult chimeras—involving either the *dy* mutant allele or the less severe *dy²ᴶ* allele—indicated that the genotype and phenotype of these tissues need not correspond. One striking case involved a muscle, all of whose fibers (or at least the vast majority) were marked as being *dy/dy;* yet the fibers were entirely normal in appearance. Apparently, then, an outside influence from *dy⁺* tissues elsewhere in the mosaic allowed for normal development and maintenance of the mutant muscle. Some of these chimeras, in fact, had most of their skeletal musculature of the *dy* genotype, suggesting that these factors were not diffusing from *dy⁺* muscles. Earlier work by Peterson (1974) had provided evidence that muscle predominantly of a *dy⁺* genotype could express dystrophic abnormalities.

These results imply that normal nerves in mouse chimeras could be the tissue that causes the muscles to be normal. Such a hypothetical influence may occur only if the normal nerves and mutant muscles are juxtaposed throughout development and postnatal stages or are together at least during some critical period during development. This "neural hypothesis" for the etiology of muscular dystrophy has, of course, not been proved by the experiments with chimeric mice. It is still possible that the *dy* gene is expressed in muscle cells and that *both* nerves and muscle would have to be mutant for degeneration of the latter to ensue. However, if the dystrophy is controlled only by the genotype of an outside influencing tissue, then there is still no evidence bearing on the genotype and phenotype of neural tissue in the mouse mosaics. Other tissues and/or a circulating factor from them could be the primary site of action of the *dy* gene.

Genetic muscular dystrophy is a wide spectrum of diseases that has been identified in many vertebrates (reviewed by Walton and Gardner-Medwin, 1974; Mendell et al., 1979; Wilson et al., 1979) and even invertebrates. Therefore, even if the case of *dy* in mouse were shown to have a neural etiology, this would not prove that a supposedly analogous disease in humans or a lower vertebrate has even the same general kind of underlying cause. However, neural etiologies have been suggested in some other forms. One experiment, which indicates the importance of analyzing a genetic disease by transplant or mosaic approaches at the earliest possible developmental stages, is that of Rathbone and colleagues (1975), in which neural tubes from chicken embryos with hereditary muscular dystrophy were transplanted into normal recipient embryos. The dystrophic neural tissue induced in the muscles of the normal host elevated levels of the enzyme thymidine kinase, a characteristic of dystrophic muscle. Thus, an early inductive effect of neural tissues on the eventual form and function of the muscles is implied in this vertebrate.

**Interactions Involved in Myelination**

Axons and Schwann cells from mutant mice with disorders of myelination in their peripheral nerves have been combined in nerve grafts to determine the etiology of the genetic neuropathies. Are they in axons, Schwann cells, or due to general systemic factors? Mutant Trembler mice (Ayers and Anderson, 1973; Low, 1977) express a dominant mutation causing severe abnormalities in peripheral nerves. All axons in the mutant are surrounded by Schwann cells, but myelin sheaths are either abnormally thin, poorly compacted, or absent. When segments of Trembler nerves were grafted into the sciatic nerves of normal mice, regenerated fibers within the graft were deficient in myelin; but in regenerated grafts of normal Schwann cells in Trembler nerves the abnormality was corrected (figure 42; Aguayo, Bray, and Perkins, 1979). The electrophysiological consequences of transplanting normal Schwann cells into Trembler nerves have also been analyzed. In intact Trembler nerves, conduction velocities are fivefold lower than normal (Low and McLeod, 1975). When segments of normal nerve were grafted into Trembler sciatic nerves, the conduction velocities across the grafted segments increased to values approaching normal levels by 6 months after the grafts were initiated (Pollard, 1978).

Mice homozygous for the recessive quaking mutation show hypomyelination in both the central and peripheral nervous systems (Sidman, Dickie, and Appel, 1964; Samorajski, Friede, and Reimer, 1970). In grafts of quaking and normal nerves performed in an analogous fashion to the Trembler experiments, it was found that grafted quaking Schwann cells failed to produce sufficient myelin when they ensheathed normal axons. However, grafts originating from normal nerves resulted in normal myelination of the segments grafted into the mutant nerves (Aguayo, Bray, and Perkins, 1979).

The findings from these two mutants indicate that the genetic neuropathies are fully expressed by transplanted Schwann cells (and their daughters) from either Trembler or quaking donors. Furthermore, the abnormalities caused by either mutation are corrected by

Proximal                  Graft                  Distal

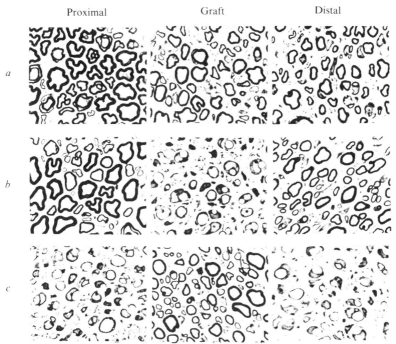

**Figure 42.**
Proximal stumps, grafted segments, and distal stump of regenerated mouse nerves 4 months after grafting. In normal-normal combinations (see *a*), the graft and distal stumps contain many regenerated myelineated fibers that resemble those in the proximal stump. When Trembler nerves are grafted into normal nerves (see *b*), the proximal and distal stumps appear normal, but fibers in the graft lack myelin or are hypomyelinated. Normal nerves grafted in Trembler nerves (see *c*) show deficient myelination in the proximal and distal stumps, whereas the grafted portion of the same nerve is normally myelinated. Phase micrographs, × 700. [Aguayo, Bray, and Perkins, 1979]

transplants of genetically normal Schwann cells. Thus, the Schwann cells themselves, not axons or systemic factors, are the "targets" of these genetic disorders, responsible for the disorders in myelination and, no doubt, the behavioral abnormalities as well (Aguayo et al., 1979).

Additional experiments have been performed on Trembler, aimed at elucidating further the details of the Schwann cell abnormalities and their interactions with axons. In the experiments involving segments of normal nerves transplanted into Trembler nerves, axon diameters became normal in the regenerated grafts but remained less than normal in the proximal and distal regions of the recipient nerves. Conversely, in regenerated grafts of Trembler Schwann cells in normal nerves, the diameters of axons were less than normal in the

grafted segment, although diameters of the same axons in the proximal and distal recipient nerves remained normal (Aguayo, Bray, and
Perkins, 1979). These observations raise the possibility that full
maturation of axonal size depends not only on factors intrinsic to that
cell (in the perikaryon or the periphery; see Aitken, 1949; Aguayo,
Peyronnard, and Bray, 1973), but also on the presence of normal
Schwann cells and myelin sheaths.

There is an increased number of Schwann cells in Trembler mice.
This increase is confined to nerves that, in normal animals, are myelinated. In the cervical sympathetic trunk (which in normal mice is
composed almost entirely of unmyelinated Remak fibers; Romine,
Bray, and Aguayo, 1976), the structure of individual Schwann cells
and the total population of nuclei of these cells are each similar in
mutant and normal mice (Aguayo, Bray, and Perkins, 1979). Thus,
there cannot be a generalized and uncontrolled disorder of the proliferation of Schwann cells in the Trembler mutant.

Still more information about Schwann cell behavior in Trembler
was obtained by grafting their unmyelinated cervical sympathetic
trunks into normal sural nerves that contained many mylelinated
fibers. When this experiment was done with normal animals, Schwann
cells in the cervical sympathetic trunk graft were induced to form myelin (Aguayo, Charron, and Bray, 1976). Although Schwann cells from
the mutant trunk ensheathed individual axons regenerating from the
normal sural nerves, they failed to produce myelin normally; and the
regenerating graft resembled a Trembler nerve (C. S. Perkins,
A. J. Aquayo, and G. M. Bray, unpublished). It would appear that
although the abnormal gene is present in all Schwann cells in Trembler
mutants, the abnormal phenotype is only expressed by those mutant
Schwann cells required to differentiate into cells that form myelin.
The basic abnormality of the Trembler mutant may therefore be
regarded as an impairment of the radial and longitudinal extension of
the plasma membrane of Schwann cells, which is necessary for normal
formation of myelin. The morphology of unmyelinated Remak fibers
is normal in Trembler nerves, presumably because the extensions of
plasma membranes by Schwann cells—required for the ensheathment
of unmyelinated axons—is less than that needed if myelination is to be
normal and complete (Aguayo et al., 1979).

## NEUROSPECIFICITY

The mechanisms by which specific nerves find their specific targets are
under intensive investigation (for recent reviews, see Edds et al., 1979;
Palka, 1979b; Anderson, Edwards, and Palka, 1980). Experimental
perturbations of input or target cells have frequently used mutations

to gain information on the events that occur as axons attempt to find their targets.

## Homeosis in Insect Development

A productive approach to the study of neurospecificity has come from recent investigations of neural connections in homeotic mutants of *Drosophila*. Homeotic mutations literally transform one body part or segment into another. It is proposed that the genes defined by these mutations have evolved along with morphological specializations of the different body segments. Examples of homeotic variants have been found in insects such as moths (Tazima, 1964), butterflies (Sibitani, 1980), mosquitos (Quinn and Craig, 1971), beetles (Englert and Bell, 1963; Daly and Sokoloff, 1965), cockroaches (Ross, 1964), houseflies (Lanna and Franco, 1961), and various species of *Drosophila* (Ouweneel, 1976). The best-studied examples of homeosis, however, are in *D. melanogaster* (Lewis, 1963; Morata and Lawrence, 1977), and it is in these organisms that neural effects have been studied (reviewed by Macagno, 1980).

Homeotic mutants offer a unique opportunity of transplantation without surgical trauma and its side effects. For example, in the Antennapedia (*Antp*) and aristapedia (*ss^a*) mutants, the antenna of the fly is transformed into a leg (figure 43). Thus, a true leg can grow out of the front of the organism's head. With such mutants it is possible to ask the following questions about the sensory projections from the homeotic organ into the CNS: Are there organ-specific nerves that tend to grow into the same part of the brain no matter what the location of the organ is on the body axis? Are there organ-specific connections in the CNS such that sensory axons of homeotic organs can make appropriate functional connections? Do the normal genes that cause a transformation of one epidermal segment into another directly affect the sensory neurons growing from each segment? Is there a normal sensory representation of body topography in homeotic mutants? These questions have been largely answered by combining newly developed neuroanatomical (and sometimes behavioral) studies with carefully devised genetic experiments.

The homeotic mutants of *Drosophila* were originally discovered and have mostly been studied with respect to their effects on the fly's external morphology. The wild-type alleles of the bithorax complex (BX-C) of genes are involved in transforming the evolutionary primitive mesothoracic-type segment of the fly into successively more posterior segments. Mutations or deletions of these genes, therefore, generally cause posterior segments to become thoracic-like (Lewis, 1963; Morata and Lawrence, 1977).

**Figure 43.**
Homeotic transformation in *Drosophila*, induced by the Antennapedia mutation. This is a scanning electron micrograph (supplied by Dr. T. C. Kaufman) of the front view of an adult *D. melanogaster*, expressing the dominant 3rd chromosomal mutation $Antp^{73b}$. Instead of two antennae developing from the relevant anterior imaginal disks, these paired tissues have been transformed into the distal portions of legs (arrows). The magnification is about × 115. [T. C. Kaufman, unpublished]

The *bx* and *pbx* mutations, when acting together, resulted in an extra pair of wings on the metathorax of the fruit fly (replacing the haltere balancer organs normally present on this posterior thoracic segment). Normally, only the mesothorax has wings; therefore, these mutations transform the metathorax to the mesothorax. The *bxd* mutation transforms the first abdominal segment into a metathorax, putting an extra pair of legs on this segment, thus transforming the insect into an eight-legged arthropod. Studies of the external morphology show that a deletion (*Df(3R)P9*) of all the genes of BX-C (hereafter called *Df*-BX-C) makes all abnormal segments of the embryo mesothoracic-like (Lewis, 1978).

Before one can properly evaluate the studies of sensory neurospecificity in such homeotic mutants, it is necessary to know whether the internal tissues, as well as the external ones, are transformed by the mutations. Unfortunately, there is some debate on this issue, as the experiments that are needed to answer the question rigorously are difficult. Ferrus and Kankel (1981) found no differences in the gross morphology of the developing CNS of several BX-C mutants, including *bx pbx / Df*-BX-C and *bxd / Df*-BX-C. Since making the mutants heterozygous with the deficiency accentuates their effect, they therefore concluded that BX-C genes probably have little or no effect on the CNS. Ghysen (1978) also proposed, from gross morphological examination of the thoracic ganglion in BC-X mutants, that only the periphery is affected by bithorax mutations; thus, only the imaginal disks and the axons they elaborate and not the CNS of the adult would be affected, thus simplifying the interpretations. Whereas this may be true of adults, Lewis (1978) showed that the homozygous deletion of the bithorax complex (*Df*-BX-C) has a dramatic effect on the development of the CNS. Among the many transformations seen in this extreme genetic variant, which survives to the end of embryogenesis, is a ventral nerve cord that seems to have retained its primitive embryonic pattern by extending to posterior segments instead of being "cephalized" in the anterior half of the animal. Parenthetically, it should be noted that this effect on the CNS may not be a true transformation whereby the ventral cord takes on evolutionary primitive characteristics (for example, annelid-like). Instead, the deletion could, in part, lead to a nonspecific derangement of embryogenesis, causing the usual foreshortening of the CNS simply to arrest. However, Campos-Ortega (unpublished) has suggested, from more detailed histological analysis of embryos deleted of the BX-C region, that the ganglia remaining in posterior segments (late in embryogenesis) are not simply present in those regions at too late a stage; rather, these ganglia appear as if they may also have been qualitatively transformed into ganglia like those of the more anterior segments.

Weaker suggestions that the CNS is altered in BX-C mutant adults come from Palka, Lawrence, and Hart (1979) and Strausfeld and Singh (1980), who noted that the metathoracic ganglia of such flies appear as if they might be somewhat enlarged (thus similar in volume to the mesothoracic ganglion). Chiarodo, Reing, and Saranchak (1971) arrived at the same conclusion from making cell counts in the appropriate mutant and normal ganglia.

Two other studies of internal tissues have assessed the effects of some BX-C mutants on larval tissues, the fat bodies (Rizki and Rizki, 1978) and muscles of the body wall (Jan and Jan, 1981). The first investigation revealed the triply mutant *bx pbx* / U*bx* larval additional (compared to wild-type) masses of anteriorly located adipose tissue. The second study showed that the *bx pbx* combination and the *bxd* mutations tend to transform the muscle patterns. Mesodermal effects of these mutations were also found in segments as posterior as the seventh abdominal, although the effects were strongest in the metathoracic and first two abdominal segments (Jan and Jan, 1981).

Effects of BX-C gene mutations on the CNS have been studied with respect to motoneurons of CNS origin. Green (1981) found that in normal animals leg motoneurons are arranged in the thoracic ganglion in segment-specific locations within each thoracic neuromere and that the axonal pathways of each group of leg motoneurons were also segment specific. The BX-C mutants *bx* and *pbx* were used in homozygous combination (*abx bx pbx / abx bx pbx*) or over a deletion to transform metathorax into mesothorax. (The, as yet, unmentioned *abx* mutation was originally designated $bx^7$ (Lewis, 1980): it causes *bx*-like transformations, but *bx* mutations and the closely linked *abx* variant are apparently not allelic.) The result was that in many such animals, the motoneurons were transformed as well as the cuticle; that is, metathoracic motoneurons were found in mesothoracic-like positions and had axonal courses typical of mesothoracic motoneurons. Green (1981) found that the level of expression of the mutations was somewhat variable but uncorrelated in the epidermis and the CNS. Thus, by motoneuron morphology, the CNS in some individuals seemed more transformed than the leg cuticle, and vice versa. These results argue rather strongly for a direct effect of BX-C genes on the CNS.

Palka and colleagues (1979), using genetic mosaics, have further suggested that there is an effect of the CNS genotype on sensory innervation of the thoracic ganglia. Large clones of $bx^3 / bx^3$ tissue were generated by induced somatic crossing over (augmented by the Minute method of Morata and Ripoll, 1975, which accentuates the size of the $bx^3 / bx^3$, Minute[+] clones); the CNS in these mosaics would almost certainly be entirely normal ($bx^3 / +$). The projection of ventral innervation from the homeotic wing to a normal CNS was similar to that seen in the thoroughly mutant case, but there were a few specific differences, one of which is that single axons and small axonal bundles in the mosaics were often seen to pass at a very ventral level (and close to the midline of the adult) into very anterior portions of the thorax, finally projecting through the prothoracic ganglion and into the animal's neck. These projections were not seen by Palka and collaborators (1979), either in wild-type flies or in the uniformly mutant cases (figure 44). Had these fibers encountered a mutant CNS, they might have terminated near their points of entry, that is, in the possibly mesothoracic-like metathorax. Instead, they contact a true metathorax, with which they normally do not make initial contact; this leads to a failure of normal termination and the qualitative misrouting to anterior parts of the animal. But the implications of these results—that bithorax mutations, in part, transform the CNS—are somewhat questionable since the $bx^3 / bx^3$ clone was of $M^+$ genotype while the rest of the fly was $M^-$. The former cells grow faster than $M^-$ ones; thus, the differences in projection specific to the mosaics may, for instance, be the result of a timing defect (Palka and Schubiger, 1980). Strausfeld and Singh (1980) have, in fact, seen the aforementioned fibers—the small bundles that take a very ventral path from the meta-

DORSAL VENTRAL

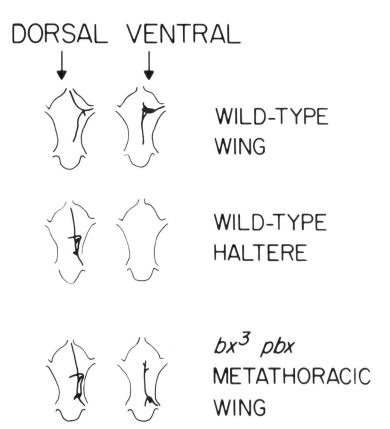

WILD-TYPE
WING

WILD-TYPE
HALTERE

$bx^3$ $pbx$
METATHORACIC
WING

**Figure 44.**
Sensory central connections affected by bithorax mutations in *D. melanogaster*. Each
picture represents a dorsal or ventral section, in the horizontal plane, through the
thoracic nervous system. The top of each picture is the anterior end of the nervous
system. Sensory axons are shown entering from the right side of each picture, either
from a wild-type wing into the mesothoracic ganglion, a wild-type haltere into the
metathoracic ganglion, or from a mutant wing ($bx^3$ $pbx$ heterozygous with an
Ultrabithorax mutation, see the text) entering the metathoracic ganglion. [Palka, Lawrence, and Hart, 1979]

thoracic entry point and then eventually go to the neck—in adults that are thoroughly mutant for bithorax alleles, that is, are not mosaic. On the other hand, since the BX-C mutants exhibit variable expression in the CNS (according to the motoneuron study), then the results of Strausfeld and Singh (1980) might sometimes be expected. Strausfeld and Singh (1980) also found in BX-C mutants that some, but certainly not all, neurons of brain origin branch in the metathoracic ganglion as they would in the mesothoracic ganglion, again suggesting an influence of these mutants on the CNS.

One last piece of evidence on CNS transformation in homeotic mutants comes from studies of the normal mesothoracic wing. If the halteres are transformed into a second pair of wings, what happens to the sensory projection from the *normal* wings? R. Burt and J. Palka (unpublished) found that these projections in such flies exhibited abnormal branching patterns in the meso- and metathorax, which implies that the CNS is directly altered.

The ensemble of results suggests that bithorax mutations do lead to transformations of portions of the CNS (at least partially and variably). It should be noted that the fat-body and muscle transformation seen in BX-C mutant larvae may not result from the direct action of these mutants on the mesodermal derivatives, but may, instead, stem from a primary change in the CNS, which then induces secondary transformations.

In a mosaic experiment concerning a different homeotic gene in *D. melanogaster,* Stocker and Lawrence (1981), using the mutant $ss^a$ (aristapedia), examined the sensory projections in homeotically transformed antennae, that is, legs on the front of the head. Since no difference was found when projections of homeotic organs were compared when the CNS was mutant (in thoroughly mutant flies) and when the CNS was wild type (in mosaics), it was concluded that $ss^a$ did not perceptibly transform the CNS.

Having raised the issue of homeotic mutants and their possible effects on the CNS, we should now examine the specific results of the investigations of the sensory connections in these mutants. Perhaps the first question is: How do the nerves from the homeotic organs grow into the CNS? Ghysen and Deak (1978) found, from irradiation or surgical disruption of larval nerves, and with several homeotic mutations, that developing adult nerves join the nearest larval nerve and appear to be guided by it into the CNS. Thus, in *Antp* (Antennapedia) and $ss^a$ (aristapedia), the axons from the homeotic leg enter the CNS by the normal antennal nerve. In cases of partial antenna-to-leg transformation, the sensory axons from both the leg and antennal parts of the organ form a single nerve typical of a normal antenna. In BX-C mutants, when the metathoracic haltere is transformed into a second wing, its nerve enters the CNS as would the haltere nerve, and not as a wing nerve. If a fourth pair of legs is homeotically added to the fly by transforming the first abdominal segment into a meta-thoracic segment, the nerve of the fourth legs enters the ganglion as would an abdominal and not a leg nerve. Palka and Schubiger (1980) and Ghysen and Janson (1980) found that the wingless ( *wg* ) mutation, which causes the wing to be transformed into a part of the thoracic cuticle called the notum, also causes the haltere nerve to enter the mesothoracic segment of the thoracic ganglion 10% of the time. This is where the wing nerve, which is obviously absent in these flies, nor-mally enters. Thus, it seems fairly clear that nerves are guided to the CNS along the closest available route and do not seek out their "appropriate" route if their organ of origin is in an ectopic location with respect to the periphery. Hence, the intrinsic genotype of an appendage does not specify where that organ will send its axons.

The most interesting results from studies of homeotic neuro-specificity concern the patterns of sensory innervation within the CNS. In the *Antp* and *ss$^a$* mutants, the antenna of the fly is, in part, transformed into a leg (see figure 43). Stimulation by sugar of a nor-mal tarsus in its correct location leads to a proboscis-extension response (Deak, 1976; Stocker, 1977). Hence, there must be connec-tions—perhaps direct or perhaps through interneurons connecting chemosensory cells on the tarsus to interneurons in the thoracic ner-vous system—that then connect with appropriate cells in the brain; the result is that the chemoreception is processed and leads to an osten-sibly normal motor output. Stimulation of a normal antenna leads to no response of the proboscis. What happens if the antenna is, in part, a leg? For both *ss$^a$* (Deak, 1976) and one allele of *Antp* ( *73b,* Stocker, 1977), stimulation of the ectopic leg led to proboscis extension. Some recent data suggest that homeotic mutations that transform the exter-nal cuticle, for example, from antenna to leg, may leave some of the sensory cells and axons developing from the relevant imaginal disk in their original state—in this case antennal (Palka and Schubiger, 1980). Yet it would seem that *Antp* must transform some of these peripheral nerve cells into a leglike state because the sense cells of a wild-type antenna in its usual location, if stimulated with sugar, do not trigger proboscis extension. If all of *Antp's* sense cells in the ectopic leg were still thoroughly antennalike, it would seem that their stimulation would not lead to the behavioral response that is observed.

The behavioral results with these antenna-into-leg mutants show that the transplanted sensory axons can still make some sort of appro-priate connections, though, of course, they enter the CNS by an ano-malous route. It is not the case that, at least for *Antp$^{73b}$,* the ectopic chemosensory axons travel all the way from the head to the thorax to find their normal targets in the thoracic ganglia. That is, the axons from the ectopic leg do appear to terminate in the brain, many of them in the parts of the CNS usually innervated by wild-type antennal nerves (Stocker et al., 1976). Chemosensory neurons from the tarsus of a homeotic leg in an $ss^a$ fly end, as do true antennal axons, in the posterior antennal center and the posterior subesophageal ganglion. However, in other regions (the antennal glomeruli and the anterior subesophageal ganglion), there are major deviations from normal antennal projections (Stocker and Lawrence, 1981). Nor do these projections correspond to regions of termination of leg sensory axons, which end solely within the thoracic ganglion (Stocker and Lawrence, 1981). Thus, some homeotic leg chemosensory neurons, forced to innervate the CNS at an abnormal place, do something anatomically novel. When the mutant sensory nerves enter the brain directly, they apparently can express some kind of intrinsic label in a dramatic way: (1) They synapse with the final targets usually innervated by the interneurons coming from the thoracic ganglia (these interneuronal brain connections are still present in the mutant because stimulation of the leg per se still leads to a normal proboscis extension); or (2) the ectopic sensory axons are able to find these interneurons relatively near their *terminals*, that is, in the brain, instead of relatively near the cell bodies in the thorax, where they are usually found by the sensory axons. The possibility in case (1) suggests that there are features of the cellular labeling that are shared by the final target neurons and the interneurons in order that the sensory axon be able to recognize either cell type.

Sensory central connections have also been intensively investigated in BX-C mutants. Ghysen (1978) and Palka and coworkers (1979) have analyzed the double mutant bithorax-postbithorax ($bx^3$ $pbx$, heterozygous with an Ultrabithorax, $Ubx$, mutation), which results in the four-wing fly discussed earlier (see also Lewis, 1963). The many sensory bristles and hairs on the normal mesothoracic wing send essentially two large groups of axons—one dorsal, one ventral—into the mesothoracic ganglion. In the main, the normal haltere sends a large dorsal bundle of axons to the metathorax from its sensory cells. In the homeotic mutant with four wings, the extra wing sends extensive dorsal and ventral inputs to the posterior portion of the ventral nervous system, which usually does not receive these two components of sensory innervation. The dorsal projection of sensory axons from the mutant wing is similar to the dorsal projection of the wild-type haltere (figure 44). In fact, the sensory structures (small campaniform sensilla), which generate this dorsal projection from the ectopic wing, are analogous to those that produce the dorsal projection from the normal haltere. Sensory neurons of these small campaniform sensilla may not even be transformed by the mutations (Palka and Schubiger, 1980). On the other hand, it is possible that the target cells themselves in the CNS are involved in determining the overall pattern of sensory innervation so that, if the small sensilla from the wing send axons to the mesothorax, the pattern is different from that observed when these sensilla, coming from a metathoracic haltere or a metathoracic wing, project their axons to the most posterior thoracic ganglion. The ectopic wing also generates a large ventral innervation of the CNS (not produced by a normal haltere); the ventral component definitely resembles the regular ventral pattern from a normal mesothoracic wing, yet the former enters the CNS far more posteriorly than normal and is essentially a mirror image of the normal mesothoracic pattern (figure 44). Part of the homeotic ventral component, a bundle of axons called the "ovoid" (Palka, Lawrence, and Hart, 1979), does *not* have its fibers seek out their normal targets in the mesothoracic ganglion; the ovoid fibers, originating from bristles on the homeotic wing, terminate posteriorly in the metathorax. Other axons in the ventral component, which originate from large campaniform sensilla of the metathoracic wing, *do* send fibers in an anterior direction. Possibly, these ventral fibers locate their normal targets in the mesothorax, in spite of having traveled an abnormally long distance to do so.

In summary, several types of behavior of sensory axons are observed as they enter the CNS via the wrong entry point, due to the "mutational transplantation" during development: (1) Certain classes of fibers can find their normal, displaced targets. (2) Other types of axons terminate at their (incorrect) entry point, but distribute themselves as if they were in their correct location. (3) Still other axonal types form a central projection that is appropriate to the normal appendage (in this case the regular metathoracic haltere), in terms of where the fibers go *and* the overall size and shape of the pattern of innervation.

It is possible that case (3), that is, axons terminating at their incorrect entry point and distributing themselves apparently according to the position of the appendage, is a result of the mutation not transforming the sensory cells. That is, external aspects of a structure would be transformed homeotically, but the sensory neuron could retain its original identity (see Palka and Schubiger, 1980). This possibility is strengthened by results from A. Ghysen and P. Santamaria (unpublished) showing that ether-induced phenocopies (for example, Capedevila and García-Bellido, 1978) of the *bx pbx* double mutant or the use of a trithorax mutation (*trx;* Ingham and Whittle, 1980) give even stronger metathoracic-to-mesothoracic transformations. In doing so, the chemical treatment or *trx* variant may transform the small campaniform sensilla of the homeotic wing. These transformed sensory cells do not make a normal haltere-like projection, but, rather, they make some elements of a winglike projection in the meso-thorax. Thus, the epidermis and the associated sensory structures may be acted upon independently. Ghysen has shown that *trx* leads to pat-chy expression of homeotic tissue on the epidermis. Studies of the sen-sory projections from the third abdominal segment, when transformed by the action of *trx,* show termination patterns in the ganglion typical of first, second, or third segment projections. Thus, there is no strict correlation between epidermal and sensory transformations, as was found with transformations of the epidermis and CNS. In addition, some BX-C mutants such as Ultra-abdominal (*Uab*) preferentially affect the epidermis and have no demonstrable effect on the sensory elements; yet other genotypes, such as a heterozygote for a deletion of BX-C and a normal chromosome, have minimal epidermal effects but pronounced neural transformations of sensory projections (A. Ghysen, unpublished). In summary, these studies leave little doubt that the pathway an axon takes in the CNS can be directly affected, independent of surrounding tissues, by the expression of homeotic genetic elements.

In *Drosophila* it has been found that the body segments are progres-sively subdivided into smaller elements called compartments (García-Bellido, Ripoll, and Morata, 1973; Crick and Lawrence, 1975). These findings, along with some results from the analysis of wing develop-ment in the bug, *Oncopeltus,* led to the suggestion that peripheral nerve cells and the axons they generate may bear specific compart-mental labels, much as the epidermal cells do (Lawrence, 1978). Fer-rus and Kankel (1981) were able to show with internal markers that certain possible compartments did not seem to exist in internal tissues, including the CNS. Their results, however, did not rule out internal compartmentalization since their data were few and not nearly all pos-sible boundary lines were analyzed.

In studying behaviors elicited by touching the mechanoreceptive structures on normal flies, Vandervorst and Ghysen (1980) found responses that seemed compartment specific. For instance, stimulated hairs in the anterior compartment of a leg gave rise to a different pat-tern of responses than did stimulation of even very nearby hairs located across the compartment boundary. Results from the wingless (*wg*) and BX-C mutations showed that ectopic bristles induce the same response that results from stimulation of wild-type bristles in their normal locations (Vandervorst and Ghysen, 1980). Thus, behav-iorally defined compartmental boundaries are observed in these trans-formed tissues.

Ghysen (1980) also found, by backfilling single sensory axons with horse radish peroxidase, that within each compartment there were slight shifts in the typical CNS projection, depending on the position of the filled sensory structure. On one hand, if neighboring sensilla were of the same modality, their central projections would be very close together, thus implying a topographical representation of the external bristle array in the CNS. If, on the other hand, neighboring sensilla were of different types, their central projections, each typical of its own modality, would be quite different, implying a different central map for each modality. When the *wg* or Contrabithorax (*Cbx*) mutations were used to eliminate some sensory structures and create new ones next to inappropriate neighbors, Ghysen (1980) found that each homeotic bristle gave a projection typical of the homologous normal bristle in its wild-type position. He further tried to disrupt topography by using the scute (*sc*) and Hairy-wing (*Hw*) mutations; *sc* eliminates thoracic bristles, and some mosaics induced with respect to the expression of this mutation had nearly all bristles eliminated. The few remaining ones gave projections typical of their precise position; there was no evidence for an expanded central projection in order to fill the projection sites in the CNS that are usually occupied but were not in this experiment because of the area denuded of bristles (Ghysen, 1980).

The *Hw* mutation adds large numbers of extra sensilla to the wing. Tracing the central pathways of these supernumerary bristles indicated that their projections also were governed by cuticular position. Extra bristles that developed in regions largely devoid of bristles gave projections qualitatively typical of bristles in that region or compartment (Ghysen, 1980), although their axons crossed compartmental boundaries while in the wing itself, on the way to the CNS (J. Palka, unpublished). These results show that the type of central projection a peripheral nerve makes is governed by (1) what type of sensory structure it originates from (a major influence), (2) which segment or compartment it belongs to (another major factor), and (3) its exact position within the compartment ("minor detail" effect).

The foregoing discussion of homeotic mutants indicates that they have been useful in generating information about the normal wiring of the *Drosophila* nervous system and the genetic and positional influences that guide it. And while some of the facts and interpretations of the effects of homeotic genes on neural specificity of peripheral-central connections or on the CNS per se are unclear, there are advantages of using these mutations to perform the relevant transplantations. The utility of the mutants compared, say, to surgical tools have been summarized (Palka, 1979a). However, nongenetic experiments, especially those done in vertebrates, have been enormously informative in defining factors underlying pattern formation in the developing nervous system (for example, Weiss, 1955). The primary interest in studying the action of homeotic genes does not center on strategies of mutational versus surgical transplantation. The most important thing to keep in mind is that the action of genes, such as Antennapedia and bithorax, are intrinsically of extreme genetic, developmental, and now neural interest, whereas surgery is of none. We may even ponder the possibility that the gene products under the control of factors such as bithorax alleles, which are likely to have striking molecular characteristics, will be identifiable in the relatively near future. One reason is that parts of the BX-C chromosomal locus have been molecularly cloned by W. Bender and colleagues (see Ashburner, 1980). This remarkable accomplishment may lead to the isolation of products that are specified by the different parts of this cluster of genes.

*Development of Isogenic Nervous Systems*
A quasi-genetic approach to the study of neurospecificity involves anatomical work on "isogenic" organisms, including both invertebrates and vertebrates (for example, Levinthal, Macagno, and Levinthal, 1975; Goodman, 1978; Macagno, 1980). A general observation is that overall patterns of connectivity, that is, how all the major sensory nerves, interneurons, and motor nerves are connected to one another, are constant among all these animals expressing the same alleles. However, several details of wiring vary; for instance, the specific number and exact location of certain fine-level connections are not the same in all animals. These results could mean that there is no genetic identity among all the neurons involved. It is actually not known whether the genome is constant in development, even whether the alleles in all cells or in all animals of an "isogenic clone" are purported to be the same. Rearranged chromosomes or transposed elements, or both, could be developmentally important sources of genetic heterogeneity (for example, McClintock, 1967; Nevers and Saedler, 1977; Green, 1980; Rubin et al., 1980). Developmentally derived genetic differences among neurons may underlie subtle wiring differences. However, there is contrary evidence from molecular studies on *Drosophila* by Potter and Thomas (1978) and nematode by Emmons and coworkers (1979). Thus, all of these nerve cells may really be isogenic, and the studies cited at the section's beginning may mean that stochastic processes operate at the level of forming the finer patterns of synaptic interactions among excitable cells. This question cannot be resolved without a real demonstration of isogenicity among neurons and other cells in these so-called clones of animals.

*Mutants of the Visual System and CNS of Mammals*
Mutations that interfere with the normal development of the mammalian nervous system can be studied at a number of levels. One can

try to discover the primary cause of the defect in molecular or cellular terms, or one can investigate secondary or compensatory changes in the nervous system that happen as consequences of the original disruption. At this second level of analysis, the use of mutations is analogous to the use of direct surgical interference. Mutations, however, can exert effects at developmental stages that are surgically inaccessible. In fact, the analysis of mammalian mutants of the CNS have pointed to specific instances in which embryonic stages of development are critical for the normal formation of certain neuronal structures. As a result, some mammalian neurobiologists interested in development have started to experiment with embryological surgery.

With the exception of humans, more neurological mutants are known for mice than for any other mammal. The reason for the preponderance of human mutations is that they make themselves known simply by arriving at a hospital or clinic. The many neurological mutations in the mouse either occurred spontaneously on long-maintained inbred stocks or were induced by radiation. About one fourth, or about 150 of the known mutations in mice, exhibit neurological or behavioral defects (Sidman, Green, and Appel, 1965). Different mutations affect many different parts of the mouse nervous system. The largest group of known neurological mutations in the mouse affect the inner ear; many also affect the cerebral hemispheres, spinal cord, and neuromuscular system. In few of these mutants has much attention been paid to changes in the neurospecificity of connections in the CNS.

*Eye-Pigment Mutants*
Mutations of the visual system are of particular interest since this system has been extensively studied in normal animals. One group of such mutations is associated with pigmentation abnormalities, the most extreme form of which is the albino mutation. The neurological defect in albinos is a reduction of the proportion of optic nerve fibers remaining ipsilateral at the optic chiasm and a commensurate increase in the proportion of contralateral fibers (Guillery, 1974). This fact was first discovered in rat and has since been found in a variety of mammals, including mouse, guinea pig, rabbit, ferret, mink, cat, tiger, and human (see Lund, 1978). In fact, the defect is expressed in all albino mammals that have so far been examined. For this reason, it is of considerable interest to understand the primary causal connection between albinism and optic nerve misrouting.

Although insufficiently analyzed, another defect associated with reduced pigmentation is in the auditory system. Both white cats (Bergsma and Brown, 1971) and human albinos (Creel et al., 1980) exhibit hearing defects that may, at least in humans, be associated with abnormal decussation of auditory fibers in the brain stem (Creel et al., 1980).

The mammalian albino locus is considered to contain the structural gene for the enzyme tyrosinase, which, among other functions, is involved in the synthesis of melanin (Foster, 1965); or this gene could involve some tyrosinase control element (Hearing, 1973). In any case, tyrosinase activity is greatly diminished in albino mutants. Thus, the following question can be posed: Is the defect in optic nerve crossing a direct result of a deficit in tyrosinase activity, or is it a secondary consequence related to reduced pigmentation of the retina? The answer seems to be the latter. We know this, first, from a study of mutations at several different loci in mink and in mouse where the pigmentation in the retinal epithelium is reduced and a misrouting of optic nerve fibers similar to the albino defect is seen (Sanderson, Guillery, and Shackelford, 1974; LaVail, Nixon, and Sidman, 1978). Mutants with reduced pigmentation elsewhere on the body, but with normal retinal pigment epithelium, have no abnormalities in their optic nerves. Second, the degree of defect in optic nerve crossing correlates reasonably well with the reduction of melanin in the pigment epithelium but little with tyrosine activity.

Although the mechanism by which eye pigmentation influences optic nerve crossing is not known, some clues are available. For instance, it has been observed that the pigment epithelium of the retina is one of the earliest structures to become pigmented in the developing embryo; here the pigment appears at a time when ganglion cells start sending out axons. Moreover, at early developmental stages the pigmentation extends for some distance beyond the eye along the optic nerve. Shortly before birth the pigment along the nerve disappears again (U. Dräger, unpublished). Perhaps the association of pigment with the developing nerves will prove to be of significance for the correlation between pigmentation of the retinal epithelium and the pattern of retinofugal projections.

Another interesting observation is that optic nerve fibers seem to grow through preformed holes, which are pigmentophobic (Silver and Sidman, 1980), in the optic stalk. Perhaps also relevant to the formation of the optic chiasm is the observation in amphibians that at least part of the retina has its embryological origin in contralateral blastomeres (Jacobson and Hirose, 1978). One of these findings may eventually shed light on the causal link between albinism and abnormal optic nerve decussation. The availability of temperature-sensitive alleles at the albino locus in mice ($c^h$, himalayan) and cats ($c^s$, Siamese) could lead to experiments designed to determine the critical period for the mutant effect if some way could be devised to manipulate the temperature of the embryo in utero.

Another promising genetic lead on the mechanism of action of a mouse albino mutation comes from a clever mosaic experiment of Guillery and coworkers (1973). These authors generated mosaic eyes through the use of an X-chromosome-autosomal translocation, for which the normal allele of the autosomal albino locus was inserted into an X-chromosome. The natural occurrence of X-inactivation—in a female embryo carrying the translocated X, a normal X, and homozygous for albino on the relevant autosome (figure 45)—leads to mosaic eyes, only parts of which will express the mutation. The question was: In the genetically mixed eyes, will an *intermediate* degree of miswiring be observed (for example, a significant, but abnormally low number of ipsilateral eye-brain connections) or will some other result obtain? The answer was the latter; the eye-brain wiring was essentially *normal*. It seemed as if a reduced but certainly nonzero level of pigment in the developing mosaic eye was sufficient to induce normal innervation from the retina to the brain. The authors then speculated on the possibility of the pigment acting as a diffusible factor, which, even when acting in a reduced amount, can developmentally direct all retinal ganglion cells to send the axons to the correct place. The alternative would have been that the "patchiness" of the mosaic retinas would cause those retinal ganglion cells adjacent to mutant pigment cells to miswire.

An additional clue to the mechanism of the albino mutation's effect on wiring at the optic chiasm comes from studies on the types of retinal ganglion cells in animals that show this miswiring. Kirk (1976) and

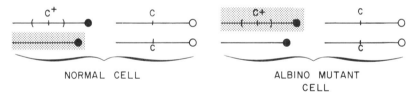

**Figure 45.**
Genetics of albino mosaicism in mouse. The gene defined by the albino mutation (c) is normally on an autosome (chromosome 7, linkage group I) in *Mus musculus*. There is a translocation in this species, resulting in the portion of this autosome containing the normal allele of the albino locus (that is, $c^+$) being inserted into the X-chromosome (Eicher, 1970); the inserted segment is delineated by parentheses. In approximately half the cells of a female mouse, one of whose X-chromosomes consists of the translocation, that chromosome will be inactivated; and the $c^+$ in the autosomal insertion may be "turned off" as a consequence. In the other half of the cells, the other, normal X-chromosome will be inactivated, leaving $c^+$ active. These inactivation events, in the two nuclei shown, are symbolized by shading. If chromosome 7 in this female is homozygous for the mutant c allele, then the cell whose nucleus is on the left will express the normal genotype regarding albinism; but the case on the right will have mutant c expression uncovered. Solid circles designate the centromeres of X-chromosomes, and open circles are the autosomal centromeres. [Guillery et al., 1973]

Cooper and Pettigrew (1979) found in Siamese cats that large ganglion cells (class Y) were most affected by the mutation. In mice, as in other vertebrates, most retinal ganglion cells are located in the ganglion cell layer of the retina, but some are found in the inner nuclear layer, in which case they are called displaced ganglion cells. In normal mice more than one fifth of all ipsilaterally projecting ganglion cells are displaced, but only about 1% of the contralaterally projecting population is displaced (Dräger and Olsen, 1980). In albino mice, overall, ganglion cells project ipsilaterally; the displaced ganglion cells are about twice as severely affected as the normal ganglion cells; that is, the proportion of displaced to normal ganglion cells projecting ipsilaterally is reduced by about half (Dräger and Olsen, 1980). The fact that (1) there is an inequality with respect to crossing at the chiasm between normal and displaced ganglion cells in wild-type mice and that (2) the albino mutation affects this inequality suggest that the defect in the mutant may be explained by an exaggeration of the normal mechanism that differentiates between these two cell types at the chiasm. For instance, the ontogenetic birthdates of these two types of cells may be different; therefore crossing at the chiasm may have to do with timing, and the defect in albino mutants may have to do with the possibility that axons first arrive at the chiasm at abnormal times.

The effects of a mutation, such as albino, in the CNS can be investigated at several levels. One can be concerned with how the lateral geniculate nucleus deals with the abnormal contralateral input in the albino mutant. This question has been most extensively studied in the Siamese cat. In the normal cat, the right lateral geniculate nucleus only receives input from the right half of each retina. The geniculate is a layered structure. The top, or most dorsal, layer, called A, receives input from the contralateral eye only; the second most dorsal layer, A1, receives ipsilateral input. In addition to this laminar organization

**Figure 46.**
Diagram to show pathways from the retina to cortex in ordinary (*A*), Boston (*B*), and Midwestern (*C*) Siamese cats. In Siamese cats, abnormal projections are shown by dashed lines, and the participating abnormal regions within the retina, LGN, and the cortex, which represent roughly the first 20° of ipsilateral visual field, are stippled. For simplicity, only connections between the right LGN and cortex have been indicated, and connections to the C-laminae and the medial normal segment of lamina A1 in the LGN have been omitted. Thus, in the LGN of both Boston and Midwestern cats, the abnormal segment of lamina A1 receives an abnormal representation of ipsilateral visual field from the contralateral eye. But at the cortical level, this abnormal representation is dealt with differently in the two varieties of Siamese cat. In Boston cats, a separate region at the anatomical 17/18 border is devoted to it, whereas in Midwestern cats, within the same region of cortex, the abnormal representation of ipsilateral visual field is superimposed upon the normal representation of the contralateral visual field. Note that the connections from peripheral retina (beyond 20°) are normal. (Shatz, 1977c]

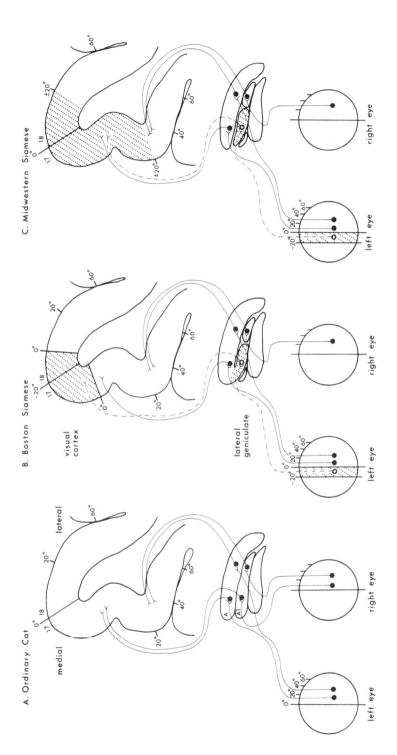

of the geniculate body, there is a retinotopic map of the world along the two-dimensional extent of each lamina, and the maps are aligned between laminas (figure 46). In the Siamese cat, the right lateral geniculate nucleus receives input from all of the right half and some (up to 20°) of the left half of the left eye, and from less than half of the right eye (Guillery and Kaas, 1971; Hubel and Wiesel, 1971; Cooper and Pettigrew, 1979). As one might expect, this situation leads to a disorganization of both the layering and the topography of the geniculate nucleus. Usually, part of lamina A1 receives input that would normally go to lamina A (see figure 46).

It is interesting to note that there are two patterns of deformation of the lateral geniculate nucleus in the Siamese cat, a Boston pattern and a Midwestern pattern, each named after the place at which they were first observed. The Boston pattern may simply be caused by a greater proportion of misrouted fibers than occurs in Midwestern-type cats (Shatz, 1977a). The differences in the geniculate between these two types of reorganization are subtle, but they lead to extensive differences in the next stage of visual processing, which occurs in the visual cortex. One more point worth noting is that, in both types of rearrangement, there is "local topography", even in the face of an overall disorganization in lamination. Based on results like these, the idea has arisen that the preservation of local topography may be a guiding influence in the development of ordered structures in the nervous system.

The abnormal geniculate in these mutants, in turn, induces abnormalities in the primary visual cortex (Hubel and Wiesel, 1971; Kaas and Guillery, 1973). The pattern of geniculocortical projection in Siamese cats depends on whether it is a Boston or Midwestern type (figure 46). In the Midwestern cortex, the geniculocortical projection is visually inappropriate (figure 46) and somehow physiologically suppressed (Kaas and Guillery, 1973). Thus, it may be that there are no higher-order consequences of the original miswiring defect in these cats.

**Figure 47.**
Two hypothetical sets of callosal connections that could exist between the anatomical 17/18 borders in each hemisphere of Boston Siamese cats. One possibility, shown in *A*, is that the two borders are connected as in common cats. (Although, of course, in common cats, the result would be to link regions serving the vertical midline representation.) This pattern, however, would be meaningless functionally since two locations in the visual field mirror, symmetrically displaced from each other by 20° to either side of the vertical midline, would be interconnected and this pattern is not found. The pattern of connections shown in *B*, which actually happens in Siamese cats, is anatomically different from that of common cats and would effectively connect two cortical sites representing identical visual field coordintes (+20°). [Shatz, 1977b]

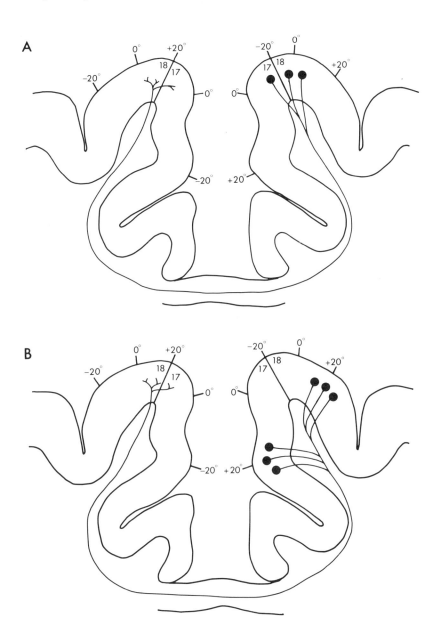

In the Boston Siamese cat, on the other hand, the entire geniculo-cortical pathway is rearranged so that the inappropriate map, as well as the contiguous appropriate map, is squeezed into the available primary cortex. As a result, the border between area 17 and area 18 no longer represents the vertical meridian, or $0°$ (Hubel and Wiesel, 1971). The abnormal projection is not suppressed in Boston Siamese cats, and thus, there are higher-order defects.

In normal cats, the corpus callosum connects the 17/18 border of each hemisphere; thus, it connects the vertical meridian of the visual field that is represented in each hemisphere, and so serves to suture the otherwise split visual world. In Siamese cats, does the corpus callosum still connect the 17/18 border (which would be inappropriately connecting $+20°$ with $-20°$) or does it still connect visually identical points in space in the two hemispheres? The latter alternative (figure 47) seems to be the accepted view. This can be demonstrated by localized injection of horseradish peroxidase (HRP; Shatz, 1977b, c), which is transported in a retrograde fashion from axon terminals to cell bodies, and by tritiated proline, which is transported anterogradely from cell bodies to axon terminals. Thus, it seems that these callosal neurons, by a mechanism that is not understood, seek out places in the contralateral visual cortex that serve the same location in the visual world.

At about the same hierarchical level of visual processing, or perhaps the next one, sixth-layer neurons in cat visual cortex project back onto the lateral geniculate nucleus. For the Boston Siamese cat there are two possibilities: the projection could be "hard-wired," that is, the 17/18 border projects to the medial edge of the geniculate as it does in normal cats but is visually inappropriate (figure 46); or the projection could be visually appropriate and therefore anatomically abnormal. Again the answer seems to be that visual appropriateness is the decisive factor (Shatz and LeVay, 1979).

Finally, it should be noted that, in the Boston Siamese cat, there does seem to be one exception to this rule of visual appropriateness: the intrahemispheric connections. In normal cats the part of area 17 that is concerned with, say, $20°$ to the right of the vertical midline projects to the visuotopically identical place in area 18 (figure 47). In Siamese cats, the region of area 17 representing the visual world $20°$ to the right of the midline projects to the region representing $20°$ to the left of the midline, also in area 17 (Shatz and LeVay, 1979). Why these visually inappropriate connections are made at the intracortical level in Boston Siamese cats, but not at the geniculocortical level, as in Midwestern Siamese cats, is unknown. In Midwestern Siamese cats, the first visually inappropriate geniculocortical connections made are physiologically suppressed. In Boston Siamese cats, inappropriate connections are made at a higher level of intrahemispheric processing, and are possibly suppressed at that stage. Could the rule be formulated, then, that a main influence in the development of connections in the visual system is visual appropriateness and that wherever this rule is violated, suppressive influences are brought into action?

Finally, it should be mentioned that the genetic basis for the difference between Boston and Midwestern cats is not understood. Occasionally, Boston-type cats are born in Midwestern colonies and vice versa. There is even the suggestion that in some cats both patterns can be present in different regions of the cortex (Cooper and Blasdel, 1979). As mentioned above, the question may be one of degree alone (Stone, Campion, and Leicester, 1978; Cooper and Pettigrew, 1979). The primary miswiring defect in Boston-type Siamese cats seems to be more severe and thus may lead to higher-order defects than those that occur in Midwestern-type Siamese cats, in which the defect may be less severe.

In human albinos there is an abnormal cortical response, demonstrated with visual-evoked potentials (VEPs), very likely due to visual-system and brain miswiring induced by pigment mutations in these primates. Components of VEPs were found to be absent in the cortical hemisphere that receives an ipsilateral, uncrossed projection (Creel, O'Donnell, and Witkop, 1978). Further work revealed two distinct and somewhat opposite types of abnormalities of VEPs in human albinos (Carroll et al., 1980). These results tentatively suggest that there are two anatomically distinct variants in the retinocortical projections of these higher albinos, just as there are in the better-analyzed Siamese cats (Carroll et al., 1980).

*Mutants with Severe Eye Defects*
Degeneration of retinal photoreceptors is a very common hereditary disease in organisms ranging from flies to man. Several distinct forms of the human affliction, called retinitis pigmentosa, are known. In laboratory mammals, retinal degeneration has mainly been studied in the rat mutant (*rdy*) and in the mouse mutant (*rd*). In *rd* mice, the degeneration starts very early, around postnatal day 8, before the photoreceptors have completed differentiation, and progresses so rapidly that by 3 weeks the retinal rods have almost all degenerated. In rat, degeneration begins later and proceeds more slowly. In both species, rods are primarily affected; cones, or at least cone somata without outer segments, may survive for the lifetime of the individual. The rest of the retina, including the bipolar, horizontal, amacrine, and ganglion cells, remains essentially unaffected by the disease. Several kinds of evidence point to the photoreceptors themselves as being primarily defective in the mouse mutant and the pigment epithelium as

being primarily defective in the rat mutant. The bulk of this evidence comes from studies of mosaic chimeras (see "Inductive Tissue Interactions"). In retinal degeneration in mice, one can still record visual responses in the superior colliculus after the rods have all degenerated (Dräger and Hubel, 1978). This response is mediated by the remaining cone cells that are devoid of outer segments.

In the deeper layers of the superior colliculus, it is possible to record somatosensory responses that are somatotopically organized and in register with the overlying visual map. Therefore, when the rods degenerate in the mouse retinal degeneration mutant, one might wonder whether the somatosensory input invades the more dorsal layers that have become largely devoid of visual input by analogy to the eyeless axolotl. One might also wonder whether the maintenance of visuotopic organization is necessary for the underlying somatotopic map; that is, in murine retinal degeneration, does the somatotopy become disorganized? The answer to both these questions is No (Dräger and Hubel, 1978). In the mouse mutant, somatosensory input to the colliculus remains normal both in layering and topography (Dräger and Hubel, 1978).

The eyeless mutant of mice has a defect in the embryonic formation of the eye. A rudimentary eye cup is formed and later resorbed. The earliest seen morphological difference between anophthalamic and normal mice is on embryonic day 10. Mesenchymal cells, which are interposed between the prospective retina and the prospective lens, become necrotic and then disappear in normal animals, but remain healthy and even keep dividing in anophthalamic mutants (Silver and Hughes, 1974). There seems to be a similar failure in the eyeless mutant of axolotl. Whether the aberrant survival of these mesenchymal cells is caused by the failure of normal eye formation, a parallel consequence of a more primary defect, or is the result of inadequate eye cup development is not known.

No optic nerve is ever formed in these animals. Surprisingly, the primary visual centers of the brain, including the lateral geniculate nucleus and the superior colliculus, are less shrunk in anophthalamic mice than in mice in which the eyes are removed at birth (Cullen and Kaiserman-Abramof, 1976). For these structures, initial innervation and subsequent denervation seem to be more destructive than never receiving innervation at all.

In the ocular retardation mutant of the mouse, the retina is thin and lacks nerve fibers. No optic nerve ever forms. The primary disturbance, while not known for certain, is thought to be a failure of vascularization through the optic fissure (Theiler et al., 1976; Silver and Robb, 1979). In both this mutant and eyeless, then, the superior colliculus never receives optic fiber input. In both of these mutants, unlike in the retinal degeneration mutant, somatosensory responses can be recorded from the most superficial layers of the colliculus, which are usually devoted exclusively to visual input. These somatosensory units at the collicular surface have unusually large receptive fields—often over one quarter to one half of the body surface—and they have a strong tendency to habituate to repeated stimuli. Somatosensory units recorded at deeper collicular levels in both mutants resemble those in normal mice: Responses are crisp and do not habituate; receptive fields are small and laid out in strict topographical order across the colliculus (Drager, 1977). It has been learned from such studies that, while rewiring can occur in the mouse superior colliculus, it appears to be somewhat selective with respect to what is rewired and the etiology of such compensatory changes.

Wiring effects have been studied in the tectum of eyeless axolotls by Gruberg and Harris (1981). The superficial neuropil of the axolotl tectum, the statum opticum, receives input from the retina, and there is evidence that at least some of this input is cholinergic (Gruberg and Greenhouse, 1973). Below the stratum opticum in normal axolotls, there is a thin layer of neuropil that receives somatosensory input as shown by electrophysiological and anatomical experiments. Correlated with this layer is a band of serotonin fluorescence. It was hypothesized by Gruberg and Solish (1978) from this kind of evidence in tiger salamanders that the somatosensory input to the tectum was serotonergic.

In eyeless axolotls, electrophysiological experiments showed that the superficial tectal neuropil was somatosensory in function. Correlated with this rearrangement of somatosensory input in eyeless, Gruberg and Harris (1981) found a matching rearrangement of the serotonin endings. Such evidence supports the notion that serotonin is used as a somatosensory neurotransmitter in the tectum. That this rearrangement is due to a secondary and not a primary defect of the eyeless genotype can be shown by removing the eyes of a normal axolotl at embryonic stages. Such genetically normal, eyeless animals show the same changes in somatosensory physiology and serotonin location in the tectum as do genetically *eyeless* animals. Moreover, grafting a normal eye onto an eyeless mutant causes the animal to show deep serotonergic fluorescence and somatosensory responses (Gruberg and Harris, 1981).

*Mutants of the CNS in Mouse*
The reeler mutation affects all cortical structures in the forebrain and cerebellum of the mouse except the olfactory bulb (see Caviness, 1977). Some aspects of the defect have been discussed in the section on "Inductive Tissue Interactions." In the current discussion, it is impor-

tant to note here, first, that there is, in reeler, a defect in the migration of cells into cortical areas. As a result, the cortex develops in an upside-down way. The cells that are usually deepest in normal mice are most superficial in reeler mice, and vice versa (Caviness, 1977). Superimposed on this upside-down cortex is a randomness that adds more confusion to the layering. The projection to the visual cortex from the lateral geniculate nucleus is also abnormal in reeler (Caviness, 1977). The final approach to layer IV, the normal terminal zone of geniculocortical afferents, is superficial to deep instead of deep to superficial. It is as if the axons were required to pass through certain layers or cell types in order to make their final termination pattern.

A question of some interest is whether reeler mice can use their visual cortex in the face of these severe abnormalities in wiring patterns and layering. Electrophysiological investigations of reeler mice show that, functionally, the cortex appears to be relatively normal (Dräger, 1977). There is normal visual topography, orientation selectivity, receptive field properties, and normal convergence of inputs from the two eyes. Not everything is normal, however. In the most superficial cells the receptive fields seem functionally more chaotic than normal, and it is difficult to find the appropriate stimulus for them (Dräger, 1977). Since, however, most cortical cells in reeler are able to establish functionally correct connections regardless of their final positions and routes of migration, the formation of connections must not be rigidly determined by these properties. Perhaps functional appropriateness or cell-to-cell affinities are more important in the establishment of intracortical connectivity.

The cerebellum, being an extremely regular and precise structure, has long been the focus of anatomical and physiological investigations. Because cerebellar defects (whether genetically or surgically induced) are generally nonlethal and cerebellar dysfunction leads to rather obvious locomotor disturbances, there are about a dozen (Mullen and Herrup, 1979) cerebellar mutants in mice, some of which have been analyzed in detail with respect to the extensive knowledge available on the normal cerebellum. The findings with these mice have provided several clues to a variety of developmental mechanisms operating in the cell differentiation, maintenance, and pattern formation of the mammalian brain. How do neurons migrate to their final locations? What happens to them if they do not? What happens to the neurons with which they usually connect? Which way do trophic influences seem to be flowing? Evidence bearing on questions like these is accumulating through the study of cerebellar mouse mutants.

The reeler mutant was first recognized for its locomotor disturbances, and only more recently studied as to its visual defects. Reelers show ataxia of gait, dystonic postures, and tremors. Gross examination reveals a reduction of the overall size of the cerebellum (see Sidman, Green, and Appel, 1965). In the cerebellar cortex of reeler, as in the visual cortex, one also finds a systematic pseudoreversed malposition of neurons. In the cerebellum of normal mice, Purkinje-cell bodies are arranged in a monolayer at the interface between the molecular layer, containing the parallel fibers and dendritic arbors of the Purkinje cells, and the granule-cell layer with the somata of these cells (figure 48). In reeler, the majority of Purkinje cells lie scattered within and below the granula cell layer.

The reason for this malposition of neurons does not lie in a disruption of events during cell division and the birth of the different cell classes in the cerebellum. In reeler, these processes are normal; nor does there seem to be any mix-up in migration from the ventricular surface according to birth date. Furthermore, granule cells migrate along glial cell processes in both normal and reeler mice, although in reeler the glial processes are sometimes oblique rather than radial (reviewed in Caviness and Rakic, 1978).

Although the cause is unclear, the earliest defect seen in reelers is in the segregation of Purkinje cells from the neurons in the deep cerebellar nuclei, an event that is usually well under way by embryonic days 13–15 (Caviness and Rakic, 1978). In reeler, the Purkinje cells remain superimposed on the deep nuclei. Then a fiber layer, which develops below the Purkinje cells in normal embryonic mice, appears superficial to them in reeler. The granule cells, which are largely superficial to this fiber zone in reeler, send out parallel fibers. However, due to their deep confinement, only a small number of Purkinje-cell dendrites find their way to the molecular layer and make contact with the granule-cell axons. From studies with chimeras made from reeler and normal mice, it is clear that neither the Purkinje nor the granule cells are primarily defective in this mutant.

Since the defect is known not to be in the Purkinje cell, the disturbance observed in their morphology must be secondary and may therefore represent generalizable defects of malpositioning and non-afferentation. Purkinje cells in reeler mutants may have one of several morphologies: (1) a normal configuration of the dendritic arbors, with a major dendritic arbor in the molecular layer, which occurs when the cell bodies happen to be in their normal position at the top of the granule cell layer; (2) cylindrical or conical configuration, which occurs when the cell body is located within or slightly below the granule layer and only a fraction of the dendritic tree is located in the molecular layer; (3) spherical or eliptical configuration, which occurs in deeply confined Purkinje cells with no extensions into the molecular layer

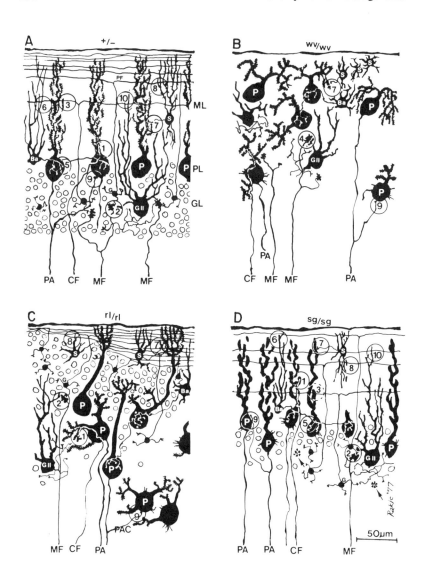

(see figure 49). Thus, it seems that axons of granule cells provide a stimulus for the proper geometric and expansive growth of the Purkinje-cell dendrites.

Certainly not all properties of Purkinje-cell morphology are dependent on parallel-fiber input. One of the remarkable autonomies of the Purkinje cell is its ability to develop small postsynaptic spines for parallel-fiber innervation even in the absence of such innervation. This is seen to be true in deeply situated Purkinje cells in reeler and weaver mutant and is also evident in animals in which the granule cells have been removed chemically, after virus infection, or by X-irradiation (Caviness and Rakic, 1978). These naked Purkinje-cell spines both form and survive in the absence of contact, yet their shapes are often somewhat bizarre. Such spines do not seem to have an obvious attraction for contacts that are from typically inappropriate cell classes.

The weaver phenotype is due to a semidomination mutation. The homozygous mutant displays cerebellar ataxia, resulting from a defect in the migration of granule cells. In homozygotes, few granule cells are able to complete a normal migration from the external molecular layer to the internal granule cell layer. The majority die at the interface between the molecular and Purkinje layer without ever sending

**Figure 48.**
Composite semischematic drawing of the neuronal arrangement and synaptic circuitry of the normal (*A*), homozygous weaver (*B*), reeler (*C*), and staggerer (*D*) cerebellum of mouse (based on Rakic, 1975). The shapes of neuronal silhouettes are drawn from Golgi preparations, and the positions of unimpregnated granule cells are outlined. All sections are oriented longitudinal to the folium and drawn at approximately the same magnification. (*A*) Normal cerebellum of a 3-week-old mouse (C57BL/6J + / + ). (*B*) Cerebellar cortex of a 3-week-old homozygous weaver mouse (C57BL/6J *wv/wv*) in a parasagittal plane where granule cells are absent. (*C*) Midsagittal outline of the 3-week-old reeler mouse (C57BL/6J *rl/rl*). The area represented in the drawing is in the area of transition between the relatively well-organized cortex where the molecular layer contains properly oriented parallel fibers (right side) and an abnormal segment of cortex with numerous granule cells situated close to the pia above the Purkinje cells (left side). (*D*) Cerebellar cortex of 2½-week-old staggerer mouse (C57BL). Many granule cells are still present at this age, although many are in the process of degeneration (arrows). Note the presence of all classes of synapses except the class between parallel fiber and Purkinje-cell dendritic spines (broken circle marked number 3): Ba, basket cell; CF, climbing fiber; G, granule cell; GII, Golgi type II cell; MF, mossy fiber; P, Purkinje cell; PA, Purkinje cell axon; PF, parallel fiber; S, stellate cell. The major classes of synapses, all identified ultrastructurally, are encircled and numbered: 1, climbing fiber to Purkinje-cell dendrite; 2, mossy fiber to granule-cell dendrite; 3, granule-cell axon (parallel fiber) to Purkinje-cell dendrite; 4, mossy fiber to Golgi type II cell dendrite; 5, basket-cell axon to Purkinje-cell soma; 6, parallel fiber to basket-cell dendrite; 7, stellate-cell axon to Purkinje-cell dendrite; 8, parallel fiber to stellate-cell dendrite; 9, Purkinje-cell axon collateral to Purkinje-cell soma; 10, parallel fiber to Golgi type II cell dendrite. [Caviness and Rakic, 1978]

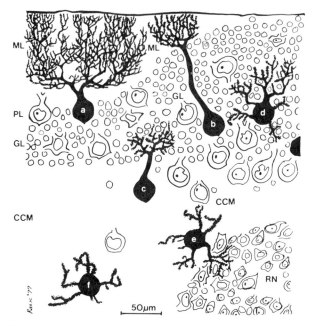

**Figure 49.**
Semidiagrammatic representation of the systematic relationship betwen the shape and size of the Purkinje cell dendritic arbor and the position of the cell in the reeler cerebellum of mouse, as visualized in sections cut transversely to the folium. To the left, a few Purkinje cells are in normal positions (PL) with respect to the molecular layer (ML) and granular layer (GL). On the right, most Purkinje cells are situated within or below the granule cells in the central cerebellar mass (CCM). Deep root nuclei (RN) lie below the Purkinje cells in the central cerebellar mass. The range of Purkinje cell shapes, based on the extensive study by Mariani and coworkers is supplemented from our own Golgi impregnation: *a,* Purkinje cell in the normal position; *b,* Purkinje cell with its soma situated in the depth of the granular layer, its single elongated dendritic shaft branches in the molecular layer; *c,* Purkinje cell with soma situated in the central cerebellar mass, its elongated dendritic shaft branches in the granular layer; *d,* Purkinje cell situated among granular cells; *e* and *f,* Purkinje cells with their somata located within the central cerebellar mass. Cells *d, e,* and *f* have multiple primary dendrites. [Caviness and Rakic, 1978]

down an axon, a process that usually precedes migration. The few granule cells that do migrate usually die.

A striking feature in this mutant is a structural abnormality in the Bergman glial fibers (Rakic and Sidman, 1973). These are radially oriented processes along which the migrating granule cells attach themselves and crawl. The irregularities in these glia consist of the underdevelopment of lamellar expansions and cytoplasmic matrices, irregularities in external contour, and some randomness in orientation.

Because of the defect in the glia, it has been hypothesized that the primary defect in weaver resides in these cells (Rakic and Sidman, 1973). Somehow the mutation impedes or disrupts the interaction between the young neurons and the Bergman glia leading to impaired migration. An alternative hypothesis puts the primary fault with the granule cells, the indication again being an impeded interaction between the granule cells and the glia (Sotelo and Changeux, 1974). Preliminary support suggesting a granule-cell focus for weaver comes from experiments by Goldowitz and Mullen (1980), who studied mosaic cerebella whose cells were labeled with respect to a marker for nuclei.

Finally, neither of these hypotheses may be correct. The defect may be in the serum of homozygous weaver mutants (Hatten, Trenkner, and Sidman, 1976; Trenkner, Hatten, and Sidman, 1976). This serum contains elevated levels of cholesterol and other lipids. When the concentration of these substances is reduced by lipid extraction, the viability of dissociated granule cells in culture is increased. Furthermore, cultures of the external granule-cell layer from weaver mice do not form the usual fasciculated processes between aggregates unless treated with the lipid-extracted serum. In fact, several of the aberrant features of the behavior of *wv* cells in culture can be corrected with such ether soluble components of serum (Trenker, Hatten, and Sidman, 1978).

As in reeler, Purkinje cells in weaver are affected in a predictable way. Without parallel fiber innervation, the dendritic branches of the Purkinje cell are oriented at random instead of fanning out in a two-dimensional array (Caviness and Rakic, 1978). And though they lack presynaptic boutons, postsynaptic spines develop anyway, but again with somewhat bizarre morphologies. Also, as in reeler, the weaver cerebellum is a good example of the general observation that synaptic specificities seem to be preserved in mutant cerebella. Wrong synaptic contacts are seldom seen. In the heterozygous weavers, ectopic granule cells in the molecular layer are contacted by mossy fibers that normally never enter the molecular layer (Caviness and Rakic, 1978).

The staggerer mutant, also showing cerebellar ataxia, is characterized by a primary defect in the Purkinje cells (Mullen and Herrup,

1979). The small postsynaptic spines, which in reeler and weaver form even in the absence of parallel fiber input, do not appear in staggerer until the third week of postnatal life, and even then are relatively few in number (Landis and Sidman, 1974). Along with the absence of spines, there appears to be an absence of normally present intramembranous particles (seen with freeze-fracture techniques) in the Purkinje-cell dendrites of staggerer mice (Landis and Reese, 1977). Purkinje cells in staggerer are thus unable to receive synaptic input from the granule cells, although they do receive normal climbing fiber input; also, they receive basket and stellate cell input (Landis and Sidman, 1978).

Somewhat surprisingly, the absence of the granule-to-Purkinje synapse is lethal to the granule cells only. This implies that there is a trophic influence of the Purkinje cell on the presynaptic neurons. It should be noted here that in reeler and weaver there is also massive granule-cell death, presumably as a result of failure to make synapses with Purkinje cells.

The Purkinje-cell-degeneration and nervous mutants are similar to staggerer in that they seem to have their primary defects in the Purkinje cells (Herrup and Mullen, 1979). In both these mutants, the Purkinje cells do form and make synaptic contact with the parallel fibers. Later, the Purkinje cells degenerate. It is interesting to note that most granule cells survive in *pcd* and nervous mutants although their synaptic targets degenerate (Caviness and Rakic, 1978), whereas in the staggerer mutant, the granule cells die, never having made contact with the Purkinje cells. Perhaps there is a critical period for the survival of granule cells during which time establishment of contact with the Purkinje cells is necessary. After this critical period, the disintegration of the postsynaptic cells may be much less traumatic for the granule cells than the situation in which no contacts had formed at all. At present, however, this is only speculation.

*Marker Mutants and Neurospecificity in Amphibians*
Many nongenetic experiments addressing questions of neurospecificity have been performed in amphibia, such as those employing rotations and transplantations of eyes. The results have shown that input cells (that is, retinal ganglion cells that send axons to the brain) can still send axons to the correct targets in the optic tectum even though they have been put in the ectopic positions (reviewed by Jacobson, 1978). Somehow, then, it appears as if the input and target cells are labeled in such a way that the growing axons can take an abnormal route to a final location containing the correct or matching label.

To augment these general findings, genetic variants have sometimes proved useful. Albino mutants are available for at least two amphibians, *Xenopus* and axolotl. Chimeric studies using the albino mutants show that the phenotype acts autonomously and is therefore an excellent marker for donor- or host-derived tissue (reviewed by Harris, 1979). Similarly, in both species, nucleolar mutations are available. When heterozygous, these mutations reduce or eliminate one of the two nucleoli per nucleus in each of the animal's cells (see Harris, 1979).

Hunt and Ide (1977) used marker mutants to study retinotectal specificity in *Xenopus*. A piece of presumptive anterior retina and pigment epithelium marked with normal pigmentation and a single nucleolus per cell from a stage 31 *Xenopus* was inverted and placed in the dorsal retina of a host marked with albinism and the normal two nucleoli per cell; both the donor and the host had developed to a similar stage. As the animal grew, the patch of retina derived from the transplant grew rapidly, forming a wedge-shaped area. There was a rough correspondence between the shape of the neural retinal patch and that of the clonally related pigment epithelium.

Hunt and colleagues have used this technique to study extensively the shape and correspondence of retinal and pigment epithelial growth in various sections of the eye (Conway, Feiock, and Hunt, 1980). In the adult, the transplanted sector of the retina projected to its appropriate part of the tectum, and it did so with a topography consistent with the transposition of the marked clone; that is, it projected to the posterior tectal pole (as anterior retina usually does), but its visual map was inverted (consistent with the surgical inversion). This experiment shows that retinal growth is radial and that positional information in the specified retina is passed on clonally or radially; yet the positional information itself is cartesian.

In another type of experiment, Gaze and coworkers (1979) reinvestigated the finding that, at an early stage of development, the host could respecify the axes of a transplanted eye. It had been claimed that, if an eye primordium were reversed left for right, up for down, or both, the resulting retinotectal map would be indistinguishable from normal as long as the operation was done early enough in embryogenesis, that is, before some critical stage (Jacobson, 1978). To pursue this further, Gaze and colleagues (1979) transplanted eyes from normal to albino *Xenopus* embryos at various times before the critical stage. In most cases, they found that, whenever the transplanted eye developed, its projections to the tectum were specified by its intrinsic orientation and the host had apparently no influence. In all other animals, they noted that part of the retinotectal map was, indeed, normal and thus independent of the orientation of the eye. It was as though this sector had been respecified by the host. In all these cases, however, the part of the eye that projected normally was of albino genotype and therefore of host, not donor, origin. In this way, Gaze and coworkers (1979) rigorously ruled out any role of the host in specifying or respecifying the axes of transplanted eyes, in whole or part, even at very early stages of development. Their work suggests that the earlier results reporting such respecification may have been due to the overgrowth of donor with host-derived tissue. Without the genetic markers, such events might have been impossible to detect.

## PHYSIOLOGICAL ACTIVITY IN DEVELOPMENT

Many investigators have entertained the notion that intracellular activity of neurons and components of intercellular communication could be involved in cell differentiation and pattern formation of nervous systems (for recent discussion and review, see, for example, McMahon, 1974; Tomkins, 1975; Changeux and Danchin, 1976; Freeman, 1977; Harris, 1981). One general idea is that already known molecules involved in carrying nerve impulses from one cell to another during postdevelopmental stages could also have a signaling function at early stages of the life cycle. Thus, one would not always have to invoke the existence of magical macromolecular morphogens. If the levels of these neurotransmitters were perturbed during development, then the differentiation of neurons, their pattern of connectivity, or at least the maintenance of normal morphology might go awry.

### Activity during Invertebrate Development

Genetic variants affecting neurotransmitter metabolism can be used to study the possible developmental importance of these molecules. Mutations in the genes for AChE (*Ace*) and CAT (*Cha*) in *Drosophila* lead to lethality at the end of embryogenesis. The CNS may have developed normally up to this stage (it appears grossly to have done so in either kind of mutant); but there is much additional neural development during larval and pupal stages (for example, White and Kankel, 1978). Tissues expressing either kind of enzymatic defect have been carried past the early lethal stage through the construction of genetic mosaics. These are, for instance, part $Ace^+$ and part $Ace^-$ (Greenspan, Finn, and Hall, 1980). The $Ace^-$ tissues that have gone through their entire development lacking AChE are structurally defective in the eventual adults. Thus, the neuropil in association with mutant cortex is abnormal (figure 50); that is, it has a compacted appearance, reduced volume, possible disorganization of axons, and in large clones (which are rare) a degenerative appearance. To interpret these effects of *Ace* mutants, it is important to determine whether there are any particular aspects of pattern that are formed under the influence of

ACh-mediated cell communication or, instead, the morphological abnormalities could all be due to deteriorations occurring after formation, or at least rather late in development.

In this regard, Hall and coworkers (1980a) have shown that the *Ace*[ts] (temperature-sensitive) mutants are most responsive to aversive effects of temperature treatments during *pupation,* obviously a late developmental stage during which the animals are apparently not exhibiting any behavior. Larvae that were undergoing development and also behaving were relatively insensitive to temperature treatments (Hall et al., 1980a). In addition, lengthy high-temperature treatments initiated after the completion of development of *Ace*[ts] flies induced morphological defects in the CNS (Hall et al., 1980a) similar to those seen in adult mosaics that had developed under the influence of nonconditional *Ace* mutations (Greenspan, Finn, and Hall, 1980).

Other mutants of physiological interest—more specifically, those related to sensory phenomena—have developmental implications as well. The gustatory mutants of Tompkins and coworkers (1979) respond abnormally to repellents such as quinine sulfate, sodium chloride, or both (see "Chemosensory Behavior"). Two of the genes defined by these mutations have been shown to influence development, once more through the use of temperature-sensitive alleles. Thus, *gusA*[Ql], which affects chemosensory behavior of larvae and adults, is cold sensitive and has a temperture-sensitive period (TSP) in embryogenesis (Tompkins, 1979). A mutation in another gene, *gusE*[N13], is also cold sensitive and affects the response of adults to quinine without affecting larval behavior. The TSP for this mutant is during the third larval instar. Other temperature-sensitive chemosensory mutants of *Drosophila* (Rodrigues and Siddiqi, 1978) have not yet been subjected to detailed temporal analysis during development. It is not known whether the activity per se of sensory cells is directly affected by any of these genes, with the developmental defects ensuing from the altered physiology. Instead, it could be that cell differentiation and/or pattern formation is affected without a specific neurophysiological etiology.

Other chemosensory mutants that affect development have been found in nematodes. At the second larval molt, if there is a scarcity of food or overcrowding, wild-type larvae may become *dauer* larvae. This is a nongrowing, nonfeeding, chemically-resistant stage of the life cycle. Dauer larvae suppress their pharyngeal pumping actions, are thermotactically reversed to normal larvae, and have an abnormally thickened epidermal cortex (for example, Riddle, 1980). Riddle (1977), Riddle, Swanson, and Albert (1981), and Albert and coworkers (1981) have isolated dauer-constitutive mutations in ten genes that cause larvae anomalously to enter the dauer stage in un-

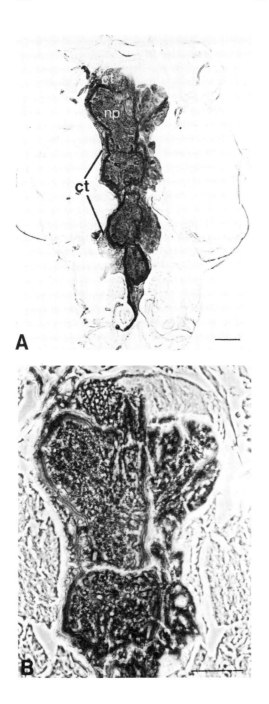

crowded, food-abundant situations. Mutations in at least twelve other genes lead to a dauer-defective phenotype in which this stage cannot form even in the absence of food. Seven of the dauer-defective variants have been found to be chemotactic mutants, lacking the normal responses to salt gradients. Three of these (one of which is an allele of the chemosensory mutant che-3) have been studied in anatomical detail; all showed different morphological defects in the anterior chemosensory cells (Albert, Brown, and Riddles, 1981; cf. Lewis and Hodgkin, 1977). These results suggest that a chemosensory signal is necessary for dauer larva formation.

If normal dauer larvae encounter a food-rich and uncrowded environment, they may transform back to the growing second-stage larvae. Is a chemosensory signal needed for this developmental process as well? Albert and coworkers (1981) were able to answer this question through the use of temperature-sensitive, dauer-constitutive mutants in combination with nonchemotactic, dauer-defective mutants. Raising the temperature causes the double-mutant combinations to transform into dauer larvae; then, lowering the temperature leaves only the dauer-defective mutation's effects. In such cases the dauer-defective mutants were unable to exit from the dauer larval stage. Thus, sensory function is needed for this transformation, as well as entry into this stage.

Another physiologically relevant gene affecting *Drosophila* development is shibire. The $shi^{ts}$ mutations, with their several effects on the functioning of excitable tissues in larva and adult, have profound effects on development when heat pulses are delivered during various stages of the ontogeny (Poodry, Hall, and Suzuki, 1973). Some of the most dramatic changes involved an induction during larval stages of, for instance, eye "scarring" (rows of undifferentiated facets) and extra sensory bristles (Dietrich and Campos-Ortega, 1980). It has been suggested that $shi^{ts}$'s mutations affect potassium channels in nerve terminals as an aspect of this gene's involvement in physiology (Salkoff and Kelly, 1978). This idea has been connected to the devlopmental relevance of $shi^{ts}$ by experiments of Salkoff and Kelly (1980) that used potassium channel blockers (aminopyridines) to

**Figure 50.**
Thoracic-abdominal nervous system of *Drosophila* in an AChE mosaic. The horizontal section was stained for AChE activity. (A) Bright-field optics showing normal AChE activity on the left side in both the cortex (ct) and neuropil (np), and AChE-null clone encompassing most of the cortex on the right side. (B) Same section under phase contrast optics at higher magnification. Mutant tissue in the cortex on the right side can be clearly seen, as well as altered neuropil morphology. Bars are 50 $\mu$m. [Hall, Greenspan, and Kankel, 1979]

induce, during the larval period, abnormalities of the eventual cuticle of wild-type adults that are quite similar to those defects seen after heat treatments of the developing mutants. Thus, it is possible that potassium channel functioning or integrity is important for certain features of cell differentiation and pattern formation. However, recall that the connections between *shi^(ts)*-induced defects in excitable cells and the mutation's possible impairment of potassium channels are likely to be quite indirect and that the action of aminopyridines on particular components of neural membranes is not thoroughly understood (Thesleff, 1980).

A single-gene mutation in another insect, the cricket, is of interest in regard to the connection between peripheral sensory activity and development of parts of the CNS. This is the filiform mutant (*fl*), isolated (and subsequently lost) by Bentley (1975). Different classes of sensory sensilla on the cerci of the animal were sequentially lost in this mutant. The mutant leads to an absence of one class of mechano-receptor hairs throughout postembryonic development (figure 51); the sensory neurons usually embedded on these hairs are thus deprived of their normal stimulation, and thus a particular interneuron in the ter-minal ganglion (the medial giant interneuron or MGI) does not receive its normal physiological input. However, the MGI is apparently inner-vated normally at the morphological level (i.e., only the peripheral hairs are affected during developmental stages by *fl*). The result of the absence of activity impinging on the MGI is that it develops into a strikingly withered or shriveled interneuron. The mutant was thus useful in uncoupling the effects exerted by sensory axons, of contact plus activity vs. contact only.

## Activity during Amphibian Development

Genetic differences between amphibian species have been used as one more way to ask neurobiological questions about the importance of known functional properties of nerve cells in certain developmental features of the nervous system. These experiments have involved chimeric individuals in which the component phenotypes differ with respect to a defined feature of neural excitation. These are the neurotoxic chimeras studied by Twitty (1937) and Harris (1980). Twitty found that parabiosis of the California newt *Taricha torosa,* which contains endogenously produced tetrodotoxin (TTX), with several species of salamanders results in the paralysis of the parabiotic salamander twin throughout embryogenesis and well into larval life. The toxin did not seem to affect the *Taricha* embryo or larva in any way. By transplanting various tissue from *Taricha* into other embryos,

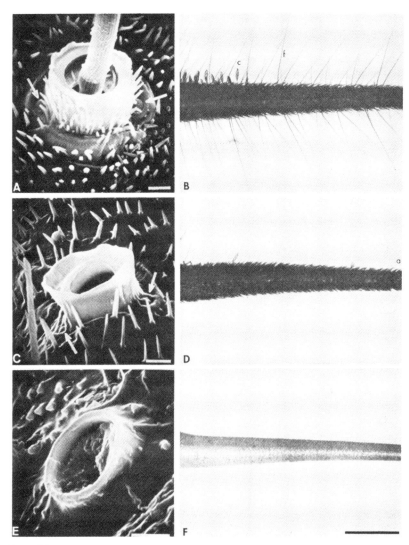

**Figure 51.**
Effects of *fl* mutation in cricket on sensilla. (*A*) Wild-type filiform hair and socket, first instar cercus (arrows indicate campaniform sensilla). (*B*) Wild-type adult cercus (*f*, filiform hair; *c*, clavate hair). (*C*) Empty filiform socket of *fl* mutant, first instar (arrows indicate campaniform sensilla). (*D*) Young *fl* mutant adult cercus (*a*, appressed hairs). Filiform and clavate hairs (save one) are missing (the long hairs at the base of the cercus are a special subclass of appressed hairs). (*E*) Empty appressed hair socket of mature adult *fl* mutant. (*F*) Mature adult *fl* mutant cercus. All hairs are missing. Calibrations: *A*, *C*, and *E*, 5 μm; *B*, *D*, and *F*, 1 mm. [Bentley, 1975]

Twitty showed that the toxin was present in all embryonic newt tissues, but concentrated in the yolk cells.

To locate the components of the neuromuscular complex affected by the toxin, Twitty (1937) carried out a series of transplantation experiments. He forced a piece of ectopic *Taricha* spinal cord to innervate an *Ambystoma* muscle in a *Taricha* host. The result of his experiments show that embryonic *Ambystoma* nerves but not muscles are paralyzed by the toxin.

By transplanting either dorsal or ventral spinal cord from *Ambystoma* to *Taricha,* Twitty (1937) next showed that all motor components of *Ambystoma* are paralyzed by the toxin, but that at least some sensory components are not. It has been shown that Rohon-Beard cells, large sensory neurons in the dorsal spinal cord of young amphibia, and some embryonic dorsal root ganglion cells first develop $Ca^{2+}$ action potentials that are TTX resistant (Baccaglini and Spitzer, 1977; Baccaglini, 1978). It would be interesting to know whether these were the cells responsible for the sensory activity in Twitty's chimeras.

One can use *Taricha* embryos as host for grafts of nervous tissues from other salamander embryos to study the role of electrical excitability in the development of the nervous system. For instance, one can ask whether the ability to make action potentials is important when two neurons are competing for the same postsynaptic target. Some experiments that address this question are outlined below.

Optic vesicles from axolotls have been transplanted as third eyes to *Taricha torosa* hosts, and vice versa (Harris, 1980). Electrophysiological studies have shown that the developing retinal ganglion cells are electrically silenced by the action of the host's tetrodotoxin. If action potentials during development are crucial in successful innervation, one might imagine that the transplanted axolotl eye competing with the host's electrically normal eye would be at a disadvantage. This does *not* seem to be the case. Neurons from an axolotl eye can extensively invade a *Taricha* tectum though the ingrowing axons are electrically silent. Results show further than an axolotl eye can invade a *Taricha* tectum so extensively that it even seems to displace the host's own retinal terminals. Thus, it is probably the case that electrical activity throughout early development is not the most critical factor in successful competitive innervation in the CNS.

These studies do not give a full picture, by any means, of the role (or lack thereof) of physiological activity in the development of the nervous system. Many studies that apply to this issue have been done without genetic techniques. The studies mentioned here, however, are of special interest because they not only are genetic manipulations but also provide some of the most clearly demonstrated developmental effects (or lack of effects) exerted by neural activity.

# 5
# CONCLUSIONS

Genetic analysis of neurobiological problems is multidisciplinary. The combination of genetics with histology, biochemistry, electrophysiology, and quantitative behavioral measurements is synergistic. Sophisticated use of these other techniques has been infused into neurogenetics in recent years and has made integral contributions to many of the successful studies. The uniqueness of the genetic approach lies partly in the kinds of technical manipulations that are possible and partly in the intrinsic relation between the disruptions caused by mutations and the elementary processes related to the nervous system. Mutations are valuable not only because they can act as "microsurgical instruments" or because "the central problem of development is the control of gene expression." Also, the genetic variants may tell us something about the manner in which the information for the nervous system is contained in the genome.

Perhaps the most tantalizing displays of this latter potential are mutants in which a characteristic is not merely deleted or blocked, but is apparently transformed into something new. Developmental examples of such mutations include those that produce neuronal structures, some of which are highly organized, in places where they are not normally found; for example, in *Drosophila,* some of the relevant mutants are Notch, lethal-X-20, several of the variants at the bithorax locus, plus other homeotic mutants, such as Antennapedia and aristapedia. Key examples in mammals include albino mutants and reeler in the mouse. There are behavioral examples as well as developmental ones: chemotactic reversal in bacteria, the nonequivalence of the optomotor phenotype in AChE mosaics of *Drosophila* to simple blindness, the vigorous courtship among *Drosophila* males as influenced by the fruitless mutation. Alterations of this sort, which are relatively specific and nonintuitive, and some of which result from the apparent *absence* of a gene product, can be of great heuristic value in attempting to understand the strategies by which the genetic instructions for behavior are implemented.

A second type of mutation derives a similar utility from phenotypes that do not completely abolish a response or process, but leave some residual function. In this regard, the phototransduction mutant in *Drosophila, norpA$^{H52}$,* is unique in that, although it has no response at high temperature, it gives a mildly abnormal response at its permissive temperature. The partial response proved to be more informative than the absence of a response for identifying an individual component of transduction that was not resolvable in the intact process. Other mutations with residual function have helped to identify parameters that are capable of quantitative alteration, for example, bacterial *che* mutants, ionic channel mutants in *Paramecium,* Shaker and transient-receptor-potential mutants in *Drosophila.* Also, in the case of the milder circadian mutants of *D. melanogaster,* their quantitative alteration in period length implicates them most convincingly in an actual clock mechanism.

Multiple alleles of genes, ranging from severely mutant to more mildly defective cases, can be of value in establishing a true link between a mutational defect and a neurological abnormality. For example, the degree of optic-system miswiring goes hand in hand with the degree of genetically induced deficit in eye pigment in different albino alleles, or even with respect to mutations in different pigment genes. Thus, the striking alterations in the organization of the CNS in, for instance, the Siamese cat, are not due to a fortuitous association of a neurological change in a strain of mammal that also happens to express an unusual pigment phenotype. When biochemical changes were found in the eyes of *Drosophila* phototransduction mutants, it was very valuable to perform the electrophoretic analysis on different mutant stocks (each carrying a separate allele of *norpA*); the changes in proteins seen in *each* strain are almost certainly not due to trivial effects of background genotype. Sometimes, mutant animals that are completely missing a behaviorally important gene, such as the dunce locus or the period locus in *Drosophila,* must and can be obtained in order to assess the consequences of a genotype that is rigorously known to be completely devoid of the relevant gene action.

Multiple alleles of recessive mutations also allow one to construct a recessive genetic defect at only one locus, through crosses involving two independently isolated alleles. Therefore, one does not have to make chromosomes homozygous that also carry mutations in other genes. This is particularly important with respect to interpretation of results from studies of (1) neurological mutants of highly inbred lines of mouse that are homozygous at most or all loci and (2) lethal mutations that must be maintained in heterozygous stocks (for example, Notch, *Ace,* or *Cha* mutants in *Drosophila*). In general, then, gratuitous effects of homozygous factors in the "genetic background" can be minimized by crossing two strains together, each of which carries an allele of the gene under study but has been inbred separately. Therefore, the resulting progeny would, at worst, be heterozygous for recessive factors at all loci except for the gene whose effects are being analyzed.

There are additional advantages to isolating several mutant alleles at a locus. Temperature sensitivity of gene expression is often attainable simply by isolating the appropriate allele of a gene, as has been found for many of the loci discussed in this book, with respect to, for instance, physiological mechanisms in *Paramecium* or *Drosophila,* muscle development in nematodes, and CNS development in *Drosophila.* Sometimes, though, achieving a temperature-sensitive tool requires the complementation of two different mutant alleles to produce a conditionally active gene product (as seen in *Ace* and dopa decarboxylase enzyme mutants in the fruit fly).

The frequency with which alleles of the same gene are repeatedly found can provide a measure of the relative number of loci that contribute to a particular behavioral or neurological phenomenon. These kinds of data exist most especially for the several mutations—and the several alleles per relevant gene—that have turned up in screens for bacterial chemotaxis variants, *Drosophila* phototransduction or chemosensory mutants, and ionic channel mutants in *Paramecium.* That extra alleles, of already identified genes, keep being found implies that the separate components of these mechanisms of cellular excitation are relatively low in number and thus tractable.

One of the most powerful uses of multiple alleles involves the analysis of interactions between mutations at different loci. For example, the study of allele-specific suppression of mutant phenotypes, with respect to the genetic control of bacterial chemotaxis, is an advanced art. Also, in *Drosophila,* allele-specific suppression in double mutants of receptor degeneration and no-receptor-potential genes is strongly suggestive of a direct link between components of a pathway. The putative interaction between a *norpA* mutant and a mutation affecting the prolonged depolarizing afterpotential has helped establish a link between phototransduction and the phenomena associated with these afterpotentials. The interactions shown between different loci affecting muscles in nematodes are likely to be similarly informative.

Studies of gene interaction may have direct molecular applications. One case is a no-action-potential ($nap^{ts}$) mutant in *Drosophila* that has been postulated to be directly defective in the regenerative sodium channels of neurons. Such channel proteins might be identifiable solely through nongenetic experiments involving toxins that bind to components of neural membranes—*if* problems of relatively "loose" and reversible binding of poisons such as TTX can be overcome. But, in addition, sodium channels are likely to be composed of more than one kind of subunit. The interaction in double mutants between the $nap^{ts}$ variant and another ($para^{ts}$) that affects nerve impulses suggests how multiple components of ionic channels, some of which may not bind to any known toxin, can be isolated.

A different kind of connection, that between distinct pathways that share common steps or components, has been established or suggested by several cases of mutants found initially by one criterion and then examined for effects on other features of excitable tissues or behavior. Some chemotactic mutants in bacteria and nematodes also affect behavioral taxes to other stimuli (for example, temperature). In *Paramecium,* pawn mutants have been used to show that avoidance reactions are controlled by mechanisms that are shared, in part, by cellular components influencing other behaviors, such as taxes toward certain chemicals. The albino mutants in mammals suggest some kind of a link between pigment synthesis in the developing eye and the formation of nerve connections that are initiated in the axonal pathways connecting the eyes to the brain. In addition, the pigment mutants reveal the striking developmental compensatory mechanisms posessed by mammals that are assembling their central nervous systems.

Other examples of mutants with strikingly different but related effects come from *Drosophila,* in which, for example, the simultaneous effects of the lethal-degenerate-disk mutant on both the imaginal cuticle and the imaginal CNS suggest that a part of a developmental pathway is common to both kinds of tissues. dunce, one of the *Drosophila* learning mutations, has been shown to affect associative conditioning and also the simpler but related phenomena of sensitization and habituation. Other learning mutants are not defective in the simpler experience-dependent behavior, suggesting a very selective and specific link between the pathways of "higher" and "lower" forms of conditioning. Similarly, the visual system mutant, optomotor-blind, isolated initially with respect to defective optomotor behavior, is also selectively impaired in pattern orientation, but not in several other visually evoked behaviors. Even more dramatic are the effects on courtship shown by *Drosophila* mutants isolated for defects in very different behaviors: an olfactory mutant, learning or memory mutants, three circadian rhythm mutants (all *per* alleles), and the optomotor mutant just mentioned. Here, the discovery of connections between the various processes has been as informative about the mutants as it has about courtship.

The experimental situations made possible by genetic technology derive their force, again, from the intrinsic nature of the genetic manipulations of an intact animal, in which nonspecific side effects are minimized. Thus, mutant and normal cells can be juxtaposed in mosaics with no surgical injury to tissues. Some of the relevant examples are pigment-photoreceptor cell interactions in the retinal dystrophy mutant of rat; Purkinje-granule cell in the staggerer mutant of mouse; eye-brain interactions in rough or Glued mutants of *Drosophila;* or ectopic nerve growth in homeotic mutants of that insect. The genetic mosaic experiments have also allowed normal and abnormal tissues to be put in a potentially interactive situation at developmental stages that are inaccessible to transplantation technique. This was especially relevant to the muscular dystrophy controversy in vertebrates. Some of the apparently conflicting findings may have to do with the fact that, using mosaics, some investigations have involved a juxtaposition of mutant and normal tissues from early developmental stages onward, while other studies did not put the two kinds of tissues together until it was, perhaps, too late to detect an influence on muscle cells by "outside" factors.

Mosaics studies with mutants of acetylcholinesterase in *Drosophila* have permitted a degree of certainty concerning the spatial distribution of enzyme blockade that is unattainable with injection or feeding of pharmacological agents. Moreover, there is always the danger in pharmacological experiments that the drugs are acting on processes other than those for which they are intended. The *Ace* mosaic experiments have also allowed local neurochemical lesions to be made, at very early developmental stages, in portions of the central nervous system that would be otherwise inaccessible. A similar degree of control over the temporal limits of an experimental perturbation is afforded by conditional alleles so that developmental or postdevelopmental effects of a genetic lesion can be assayed. Finally, the neurotransmitter mutants and mosaics have allowed one to make deep or severe lesions in neural functioning, but with the animal left otherwise intact. Thus, it has been possible to determine in a meaningful way whether the physiologically disrupted individual can carry out a behavior. Even the most sophisticated fixed-action patterns, involving complex movements of the whole animal and its appendages, are amenable to study because there will not necessarily be a general debilitation that could result from surgical intervention, injection of a drug through peripheral into central areas, or from being tethered by impaled electrodes.

In general, the studies on the genetics of neurotransmitters represent something of a departure from the usual neurogenetic analysis. Transmitter neurogenetics begins with a search for the relevant mutations based on specific neurochemical or enzymatic impairments, instead of the more usual behavioral criteria. Subsequently, *Ace* mutants in *Drosophila,* and the other variants in this organism and the nematode, have been used to perturb the nervous system, including its control of behavior. The purpose of using these mutational perturbations was not only to achieve basic background information on the function of these transmitter molecules, but also to allow a high degree of manipulability of transmitter metabolism in order to ask rather broad questions about the biology of transmitter function. The purpose was not, for instance, to prove that acetylcholine is a neuro-transmitter in *Drosophila*.

The manipulability provided by transmitter mutations or variants affecting nerve impulses (in *Drosophila* and newt) allows one to augment greatly the kinds of approaches available from pharmacology. Thus, even though these neurochemical genetic approaches may not lead to the discovery of, as yet, unknown molecules of neurobiological interest, they are still valuable as an adjunct to the older and more common modes of analyzing these functions.

The ability of some mutations to eliminate specific parts or classes of cells represents another form of microsurgery. The mutant strain of bacteria without flagella and with polyhooks was extremely useful in demonstrating that these cells move specifically by rotating their flagella. *Drosophila* mutants of the eye, such as receptor-degeneration, outer-rhabdomeres-absent, and sevenless, have been enormously useful as tools to delete certain cell types of the retina in order to uncover particular functions within different kinds of retinula cells and to yield information on the different behavioral pathways into which various kinds of photoreceptor cells feed. The optomotor-blind mutation performs an apparently selective elimination of cells in the lobula plate of *Drosophila,* and the neural control of later steps in visual pathways was thus probed with this mutant. The cell lineage mutants of the nematode, by deleting specific classes of cells, have provided developmental information on the regulative capacity of the nervous system.

The specificity of effects exerted by many neurological mutations might cause one to imagine that there are "genes for the nervous system" or even "genes for specific parts of the nervous system." Critics of the genetic approach to neurobiology have sometimes reacted strenuously to these possibilities since they believe that nearly all developmental mutations will be pleiotropic in their actions, even if only one tissue appears to be overtly abnormal in a given mutant. Hence, studies of neurodevelopmental mutants are claimed to be misguided with respect to eventually revealing mechanisms of neural assembly. Such experiments might be regarded as analogous to throwing a wrench into a complex machine, fouling its function generally, and then trying somehow to understand the way it works.

We wish to stress in the strongest terms that these criticisms are the feeblest kind of "straw man" attacks on neurogenetics. For no neurogeneticist has claimed to have discovered neural-specific or cell-type-specific genes through the isolation of their mutants. Most investigators fully realize that such mutations may affect other tissues, though not obviously, and that additional mutant alleles of the genes could be more widespread in their overt effects or even be lethal. Indeed, additional variants of these genes have been isolated in many cases, some-times through deliberately extended mutagenesis work, including the construction of deletions of certain loci. Therefore, the spectrum of possible defects induced by mutations in a neurobiologically inter-esting gene can be assessed through experiments, not polemics. And if only certain tissues appear to be abnormal even in the face of the "worst" possible defect in gene action (for example, a homozygous deletion), one has other ways eventually to determine whether the gene really is expressed in many parts of the organism. The molecular clon-ing of DNA from neurally important genes (already achieved or well under way for several such genetic loci) will allow further tests of expression of these genes; it may turn out that messenger RNA tran-scripts from the cloned DNA of certain loci will be detectable in only a limited range of tissues. On the other hand, it is possible that these techniques will reveal a lack of tissue specificity for expression of nearly all so-called nerve-specific genes. This would imply that, in fact, at least some of the variants associated with such loci will be demonstrably pleiotropic. We believe that these facts and principles, and the additional information they promise to produce, indicate that one must *expand and deepen the genetic approach to neurobiology.* We cannot decide whether there are or are not genes for the nervous system merely by assertion. And as we continue to analyze existing mutants neurobiologically, we realize, as well, that it is essential to apply a great deal of further genetic and molecular technology to allow an eventual resolution of these mutants—quite the opposite of an abandonment of the genetic strategies.

As answers to questions on nerve-specific genes continue to be obtained, the exquisite selectivity of effects of certain mutant alleles will be increasingly used in behavioral as well as developmental experi-ments. Further examples of mutants as tools include the use of condi-tional paralytic variants to assay the role of female activity in *Droso-phila* courtship; blind or olfactory mutants used to define the roles of these sensory inputs in the courtship of this fly; "endocrine behav-ioral" mutants used to study components of reproduction in the axolotl and the mouse; and mouse mutants affecting certain cell classes in the brain that are used to ask questions about cell-specific neurochemical functions. The latter type of mouse mutants, selec-tively deleting certain cell types within the cerebellum, have also been important in revealing additional defects of neural assembly or maintenance that can apparently "cascade" from a defect in, or absence of, any one of several different classes of cells (depending on the mutant used). In the cricket, the selective elimination by a mutant of certain sensory hairs was important in discriminating between the effects on interneuronal structure of sensory-central connectivity per se and the additional influence of physiological communication between sense cells and their ganglionic targets.

Apart from the intriguing phenotypes that many of the neurological mutants exhibit and the clever manipulations that have been described, for example, interacting genes, conditional alleles, and genetic mosaics, there remains a central role for formal chromosome genetics. The production and manipulation of chromosomal rearrangements has been mandatory for gene cloning of factors controlling chemotaxis in bacteria and the resultant molecular information on the products of such genes. The use of rearranged chromosomes has also been critical for many mosaic studies, which have, for instance, required translocation of autosomal loci into a sex chromosome (in *Drosophila* or mouse). Other kinds of chromosome rearrangements have led to the development of the powerful segmental aneuploidy techniques that have launched several neurogenetic analyses on the action of known gene products. Such genetic sophistication, aimed at mosaic construction, gene localization, and mutation isolation, is not limited to *Drosophila* because genetic techniques that can be used for producing mosaics and identifying genes in nematodes are now coming to the fore.

The use of genetic deletions can uncover important new mutant phenotypes, such as (1) those resulting from deletion of the entire bithorax complex in *Drosophila;* (2) deletion of parts of the scute-bristle region in that organism; and (3) deletion of part of a myosin gene in the nematode and suppression by a deletion of a particular muscle mutant in this roundworm. Deletions can also facilitate careful genetic mapping, such as that which led to the striking correlation of a learning mutant in *Drosophila* with a genetic locus controlling a cyclic nucleotide phosphodiesterase.

In conclusion, advancements in our understanding of neurogenetics will require not only detailed studies drawing on a host of established techniques of neurobiology, but also continuing exploitation of the sophisticated genetic principles and techniques available in most of the organisms described in this report, animals in which studies of genetics and of excitable cells and tissues are being so intimately merged.

# ABBREVIATIONS

| | |
|---|---|
| ACh | acetylcholine |
| AChE | acetylcholinesterase |
| 4-AP | 4-aminopyridine |
| $\alpha$-BT | $\alpha$-bungarotoxin |
| cAMP | cyclic adenosine monophosphate |
| CAT | cholineacetyltransferase |
| CNS | central nervous system |
| cGMP | cyclic guanosine monophosphate |
| ERG | electroretinogram |
| FIF | formaldehyde-induced fluorescence |
| GABA | $\gamma$-aminobutyric acid |
| GnRH | gonadotropin-releasing hormone |
| HGPRT | hypoxanthine-guanine phosphoribosyltransferase |
| MAO | monoamine oxidase |
| MCP | methyl-accepting chemotaxis proteins (of bacteria) |
| MGI | median giant interneuron (of cricket) |
| P | precursor cell (in developing nematode) |
| PDA | prolonged depolarizing afterpotential |
| STX | saxitoxin |
| TEA | tetraethylammonium |
| TSP | temperature-sensitive period |
| TTX | tetradotoxin |
| UV | ultraviolet |
| VEP | visual-evoked potentials |
| VNS | ventral nervous system |

# BIBLIOGRAPHY

Aceves-Piña, E. O., and Quinn, W. G. (1979): Learning in normal and mutant *Drosophila* larvae. *Science* 206:93–96.

Adams, D. J., Smith, S. J., and Thompson, H. (1980): Ionic currents in molluscan soma. *Ann. Rev. Neurosci.* 3:141–161.

Adler, J. (1969): Chemoreceptors in bacteria. *Science* 166:1588–1597.

Adler, J. (1975): Chemotaxis in bacteria. *Ann. Rev. Biochem.* 44:341–356.

Adler, J. (1976): The sensing of chemicals by bacteria. *Sci. Am.* 234(4):40–47.

Adler, J., and Dahl, M. M. (1967): A method for measuring the motility of bacteria and for comparing random and non-random motility. *J. Gen. Microbiol.* 46:161–173.

Adoutte, A., Ramanathan, R., Lewis, R. M., Dute, R. D., Ling, K.-Y., Kung, C., and Nelson, D. L. (1980): Biochemical studies of the excitable membrane of *Paramecium tetraurelia*. III. Proteins of cilia and ciliary membranes. *J. Cell Biol.* 84:717–738.

Aguayo, A. J., Attiwell, M., Trecarten, J., Perkins, S., and Bray, G. M. (1977): Abnormal myelination in transplanted Trembler mouse Schwann cells. *Nature* 265:73–75.

Aguayo, A. J., Bray, G. M. and Perkins, C. S. (1979): Axon-Schwann cell relationships in neuropathies of mutant mice. *Ann. N.Y. Acad. Sci.* 317:512–533.

Aguayo, A. J., Bray, G. M., Perkins, C. S., and Duncan, I. D. (1979): Axon-sheath cell interactions in peripheral and central nervous system transplants. In *Society for Neuroscience Symposia,* Vol. IV. *Aspects of Developmental Neurobiology.* Ferrendelli, J. A., ed. Bethesda: Society for Neuroscience, pp. 361–383.

Aguayo, A. J., Charron, L., and Bray, G. M. (1976): Potential of Schwann cells from unmyelinated nerves to produce myelin: A quantitative ultrastructural and autoradiographic study. *J. Neurocytol.* 5:565–573.

Aguayo, A. J., Peyronnard, J. M., and Bray, G. M. (1973): A quantitative ultrastructural study of regeneration from isolated proximal stumps of transected unmyelinated nerves. *J. Neuropathol. Exp. Neurol.* 32:256–270.

Aitken, J. T. (1949): The effect of peripheral connexions on the maturation of regenerating nerve fibres. *J. Anat.* 83:32–43.

Aksamit, R. R., and Koshland, D. E., Jr. (1974): Identification of the ribose binding protein as the receptor for ribose chemotaxis in *Salmonella typhimurium. Biochemistry* 13:4473–4478.

Alawi, A. A., Jennings, V., Grossfield, J., and Pak, W. L. (1972): Phototransduction mutants of *Drosophila melanogaster.* In *The Visual System: Neurophysiology, Biophysics, and Their Clinical Applications.* Arden, G. B., ed. New York: Plenum Press, pp. 1–21.

Alawi, A. A., and Pak, W. L. (1971): On-transient of insect electroretinogram: Its cellular origin. *Science* 172:1055–1057.

Albert, P. S., Brown, J., and Riddle, D. L. (1981): Sensory control of dauer larva formation in *Caenorhabditis elegans. J. Comp. Neurol.* 198:435–451.

Albertson, D. G., Sulston, J. E., and White, J. G. (1978): Cell cycling and DNA replication in a mutant blocked in cell division in the nematode *Caenorhabditis elegans. Dev. Biol.* 63:165–178.

Alnaes, E., and Rahamimoff, R. (1975): On the role of mitochondria in transmitter release from motor nerve terminals. *J. Physiol.* 248–306.

Anderson, H., Edwards, J. S., and Palka, J. (1980): Developmental neurobiology of invertebrates. *Ann. Rev. Neurosci.* 3:97–139.

Armstrong, C. M., and Binstock, L. (1965): Anomalous rectification in the squid giant axon injected with tetraethylammonium chloride. *J. Gen. Physiol.* 48:859–872.

Armstrong, J. B., Adler, J., and Dahl, M. M. (1967): Nonchemotactic mutants of *Escherichia coli. J. Bacteriol.* 93:390–398.

Arnett, D. W. (1972): Spatial and temporal integration properties of units in first optic ganglion of Dipterans. *J. Neurophysiol.* 35:429–444.

Arnold, A. P. (1980): Sexual differences in the brain. *Am. Sci.* 68:165–173.

Ashburner, M.(1980): *Drosophila* at kolymbari. *Nature* 288–540.

Aswad, D., and Koshland, D. E., Jr. (1974): Role of methionine in bacterial chemotaxis. *J. Bacterioi.* 118:640–645.

Atwood, H. L., Swenarchuk, L. W., and Gruenwald, C. R. (1975):

Long-term synaptic facilitation during sodium accumulation in nerve terminals. *Brain Res.* 100:198–204.

Averhoff, W. W., and Richardson, R. H. (1974): Pheromonal control of mating patterns in *Drosophila melanogaster. Behav. Genet.* 4:207–225.

Ayers, M. M., and Anderson, R. M. (1973): Onion bulb neuropathy in the trembler mouse: A model of hypertrophic interstitial neuropathy (Dejerine-Sottas) in man. *Acta Neuropathol.* 25:54–70.

Baccaglini, P. I. (1978): Action potentials of embryonic dorsal root ganglion neurones in *Xenopus* tadpoles. *J. Physiol.* 283:585–604.

Baccaglini, P. I., and Spitzer, N. C. (1977): Developmental changes in the inward current of the action potential of Rohon-Beard neurones. *J. Physiol.* 271:93–117.

Bachman, B. J., Low, K. B., and Taylor, A. L. (1976): Recalibrated linkage map of *Escherichia coli* K12. *Bacteriol. Rev.* 40:116–167.

Balinsky, B. I. (1958): On the factors controlling the size of the brain and eyes in Anuran embryos. *J. Exp. Zool.* 139:403–441.

Ballard, W. W. (1939): Mutual size regulation between eyeball and lens in Amblystoma, studied by means of heteroplastic transplantation. *J. Exp. Zool.* 81:261–285.

Barchas, J. D., Elliott, C. R., and Berger, P. A. (1978): Biogenic amine hypotheses of schizophrenia. In *The Nature of Schizophrenia: New Approaches to Research and Treatment.* Wynne, L. C., Matthysse, S., and Cromwell, R., eds. New York: John Wiley, pp. 126–142.

Barnett, A. (1966): A circadian rhythm of mating type reversals in *Paramecium multimicronucleatum,* syngem 2, and its genetic control. *J. Cell Physiol.* 67:239–270.

Barnett, A. (1969): Cell division: A second circadian clock system in *Paramecium multimicronucleatum. Science* 164:1417–1418.

Barth, R. H., and Lester, L. J. (1973): Neuro-hormonal control of sexual behavior in insects. *Ann. Rev. Entomol.* 18:445–472.

Beadle, G. W., and Tatum, E. L. (1941): Genetic control of biochemical reactions in *Neurospora. Proc. Nat. Acad. Sci.* 27:499–506.

Beckman, C. (1970): *sk:* stuck: *Drosophila Inf. Serv.* 45–36.

Bentley, D. (1975): Single gene cricket mutations: effects on behavior, sensilla, sensory neurons, and identified interneurons. *Science* 187:760–764.

Benzer, S. (1967): Behavioral mutants of *Drosophila* isolated by counter-current distribution. *Proc. Nat. Acad. Sci.* 58:1112-1119.

Benzer, S. (1971): From the gene to behavior. *JAMA* 218:1015-1022.

Benzer, S. (1973): Genetic dissection of behavior. *Sci. Am.* 229(6): 24-37.

Berg, H. C. (1975): Chemotaxis in bacteria. *Ann. Rev. Biophys. Bioeng.* 4:119-136.

Berg, H. C., and Brown, D. A. (1972): Chemotaxis in *Escherichia coli* analysed by three-dimensional tracking. *Nature* 239:500-504.

Bergsma, D. R., and Brown, K. S. (1971): White fur, blue eyes and deafness in the domestic cat. *J. Hered.* 62:171-185.

Berman, N., and Hunt, R. K. (1975): Visual projections to the optic tecta in *Xenopus* after partial extirpation of the embryonic eye. *J. Comp. Neurol.* 162:23-42.

Bicker, G., and Reichert, H. (1978): Visual learning in a photoreceptor degeneration mutant of *Drosophila melanogaster*. *J. Comp. Physiol.* 127:29-38.

Bishop, L. G., and Keehn, D. G. (1966): Two types of neurones sensitive to motion in the optic lobe of the fly. *Nature* 212:1374-1376.

Black, I. B., and Geen, S. C. (1974): Inhibition of the biochemical and morphological maturation of adrenergic neurons by nicotinic receptor blockade. *J. Neurochem.* 22:301-306.

Black, R. A., Hobson, A. C., and Adler, J. (1980): Involvement of cyclic GMP in intracellular signalling in the chemotactic response of *Escherichia coli*. *Proc. Nat. Acad. Sci.* 77:3879-3883.

Bon, S., Vigny, M., and Massoulié, J. (1979): Asymmetric and globular forms of acetylcholinesterase in mammals and birds. *Proc. Nat. Acad. Sci.* 76:2546-2550.

Booker, R., and Quinn, W. G. (1981): Conditioning of leg positioning in normal and mutant *Drosophila*. *Proc. Nat. Acad. Sci.* 78.

Boos, W. (1969): The galactose binding protein and its relationship to the β-methylgalactoside permease from *Eschetichia coli*. *Eur. J. Biochem.* 10:66-73.

Bovet, D., Bovet-Nitti, R., and Oliverio, A. (1969): Genetic aspects of learning and memory in mice. *Science* 163:139-149.

Boyse, E. A., and Cantor, H. (1978): Immunogenetic aspects of biologic communication: A hypothesis of evolution by program duplication. In *The Molecular Basis of Cell-Cell Interaction*. Lerner, R. A., and Bergsma, D., eds. New York: Alan R. Liss, pp. 249-283.

Bradley, W. G., and Jenkinson, M. (1973): Abnormalities of peripheral nerves in murine muscular dystrophy. *J. Neurol. Sci.* 18:227–247.

Braitenberg, V. (1972): Periodic structures and structural gradients in the visual ganglia of the fly. In *Information Processing in the Visual System of Arthropods*. Wehner, R., ed. Berlin: Springer-Verlag, pp. 3–15.

Bray, G. A., and York, D. A. (1971): Genetically transmitted obesity in rodents. *Physiol. Rev.* 51:598–646.

Breakefield, X. O., Edelstein, S. B., and Castro Costa, M. R. (1979): Genetic analysis of neurotransmitter metabolism in cell culture: Studies on the Lesch-Nyhan syndrome. In *Neurogenetics: Genetic Approaches to the Nervous System*. Breakefield, X. O., editor-in-chief. New York: Elsevier, pp. 197–234.

Brehm, P., and Eckert, R. (1978): CA-dependent inactivation of calcium conductance in *Paramecium*. *Soc. Neurosci. Abstr.* 4:234.

Brenner, S. (1974): The genetics of *Caenorhabditis elegans*. *Genetics* 77:71–94.

Brown, D. A., and Berg, H. C. (1974): Temporal stimulation of chemotaxis in *Escherichia coli*. *Proc. Nat. Acad. Sci.* 71:1388–1392.

Bruce, V. G. (1972): Mutants of the biological clock in *Chlamydomonas reinhardi*. *Genetics* 70:537–548.

Bruce, V. G., and Bruce, N. C. (1978): Diploids of clock mutants of *Chlamydomonas reinhardi*. *Genetics* 89:225–233.

Brun, R. B. (1978): Experimental analysis of the eyeless mutant in the Mexican axoloti (*Ambystoma mexicanum*). *Am. Zool.* 18:273–279.

Bryant, P. J., and Zornetzer, M. (1973): Mosaic analysis of lethal mutations in Drosophila. *Genetics* 75:623–637.

Buchner, E. (1976): Elementary movement detectors in an insect visual system. *Biol. Cybernet.* 24:85–101.

Buchner, E., and Buchner, S. (1980): Mapping stimulus-induced nervous activity in small brains by [³H]2-deoxy-D-glucose. *Cell Tiss. Res.* 211:51–64.

Buchner, E., Buchner, S., and Hengstenberg, R. (1979): 2-Deoxy-D-glucose maps movement-specific nervous activity in the second visual ganglion of *Drosophila*. *Science* 205:687–688.

Burnet, B., Connolly, K., and Mallinson, M. (1974): Activity and sexual behavior of neurological mutants in *Drosophila melanogaster*. *Behav. Genet.* 4:227–235.

Burnet, B., Eastwood, L., and Connolly, K. (1977): The courtship song of male Drosophila lacking aristae. *Animal Behav.* 25:460-464.

Buzin, C. H., Dewhurst, S. A., and Seecof, R. L. (1978): Temperature sensitivity of muscle and neuron differentiation in embryonic cell cultures from the *Drosophila* mutant, *shibire*[ts1]. *Dev. Biol.* 66: 442-456.

Byers, D., Davis, R. L., and Kiger, J. A., Jr. (1981): Defect in cyclic AMP phosphodiesterase due to the dunce mutation of learning in *Drosophila melanogaster. Nature* 289:79-81.

Byrne, B. J., and Byrne, B. C. (1978): An ultrastructural correlate of the membrane mutant "paranoiac" in *Paramecium. Science* 199: 1091-1093.

Calladine, C. R. (1974): Bacteria can swim without rotating flagellar filaments. *Nature* 249:385.

Campos-Ortega, J. A. (1980): On compound eye development in *Drosophila melanogaster.* In *Current Topics in Developmental Biology,* Vol. 15. *Neural Development, Part 1. Emergence of Specificity in Neural Histogenesis.* Hunt, R. K., ed. New York: Academic Press, pp. 347-371.

Campos-Ortega, J. A., and Hofbauer, A. (1977): Cell clones and pattern formation: On the lineage of photoreceptor cells in the compound eye of *Drosophila. Wilhelm Roux' Arch. Entwicklungsmech. Org.* 181:227-245.

Campos-Ortega, J. A., and Jimenéz, F. (1980): The effect of X-chromosome defiencies on neurogenesis in *Drosophila.* In *Development and Neurobiology of Drosophila.* Siddiqi, O., et al., eds. New York: Plenum Press, pp. 201-222.

Campos-Ortega, J. A., Jurgens, G., Hofbauer, A. (1979): Cell clones and pattern formation: Studies on *Sevenless,* a mutant of *Drosophila melanogaster. Wilhelm Roux' Arch. Entwicklungsmech. Org.* 186: 27-50.

Campos-Ortega, J. A., and Strausfeld, N. S. (1972): The columnar organization of the second synaptic region of the visual system of *Musca domestica* L. I. Receptor terminals in the medulla. *Z. Zellforsch.* 124:561-585.

Capdevila, M. P., and García-Bellido, A. (1978): Phenocopies of bithorax mutants: genetic and developmental analyses. *Wilhelm Roux' Arch. Dev. Biol.* 185:105-126.

Carroll, W. M., Jay, B. S., McDonald, W. I., and Halliday, A. M. (1980): Two distinct patterns of visual evoked response asymmetry in human albinism. *Nature* 286:604-606.

Carter-Dawson, L. D., LaVail, M. M., and Sidman, R. L. (1978): Differential effect of the rd mutation on rods and cones in the mouse retina. *Invest. Ophthalmol. Vis. Sci.* 17:489–498.

Caviness, V. S., Jr. (1977): Reeler mutant mouse: A genetic experiment in developing mammalian cortex. In *Society for Neuroscience Symposia,* Vol. II. *Approaches to the Cell Biology of Neurons.* Cowan, W. M., and Ferrendelli, J. A., eds. Bethesda: Society for Neuroscience, pp. 27–46.

Caviness, V. S., Jr., and Rakic, P. (1978): Mechanisms of cortical development: A view from mutations in mice. *Ann. Rev. Neurosci.* 1:297–326.

Chalfie, M., Horvitz, H. R., and Sulston, J. E. (1981): Mutations that lead to reiterations in the cell lineages of C. elegans. *Cell* 24:59–69.

Chance, B., Estabrook, R. W., and Ghosh, A. (1964): Dampened sinusoidal oscillations of cytoplasmic reduced pyridine nucleotide in yeast cells. *Proc. Nat. Acad. Sci.* 51:1244–1251.

Chance, B., Pye, E. K., Ghosh, A. K., and Hess, B. (eds.) (1973): *Biological and Biochemical Oscillators.* New York: Academic Press.

Chang, S. Y., and Kung, C. (1976): Selection and analysis of a mutant *Paramecium tetraurelia* lacking behavioral response to tetraethylammonium. *Genet. Res.* 27:91–107.

Changeux, J.-P., and Danchin, A. (1976): Selective stabilisation of developing synapses as a mechanism for the specification of neuronal networks. *Nature* 264:705–712.

Chelsky, D., and Dahlquist, F. W. (1980): Structured studies of methyl-accepting chemotaxis proteins in *Escherichia coli:* Evidence for multiple methylation sites. *Proc. Nat. Acad. Sci.* 77:2434–2438.

Chevais, S. (1937): Sur la structure des yeux implantés de *Drosophila melanogaster. Arch. Anat. Microsc. Morphol. Exp.* 33:107–112.

Chiarodo, A., Reing, C. M., Jr. and Saranchak, H. (1971): On neurogenetic relations in *Drosophila melanogaster. J. Exp. Zool.* 178: 325–330.

Childs, B. (1972): Genetic analysis of human behavior. *Ann. Rev. Med.* 23:378–406.

Childs, B., Finucci, J. M., Preston, M. S., and Pulver, A. (1976): Human behavior genetics. *Adv. Human Genet.* 7:57–97.

Ciaranello, R. D., Lipsky, A., and Axelrod, J. (1974): Association between fighting behavior and catecholamine biosynthetic enzyme activity in two inbred mouse sublines. *Proc. Nat. Acad. Sci.* 71: 3006-3008.

Clark, A. M., and Egen, R. C. (1975): Behavior of gynandromorphs of the wasp *Habrobracon juglandis*. *Dev. Biol.* 45:251-259.

Collins, A. L., and Stocker, B. A. D. (1976): *Salmonella typhimurium* mutants generally defective in chemotaxis. *J. Bacteriol.* 128:754-765.

Cone, R. A. (1973): The internal transmitter model for visual excitation: Some quantitative implications. In *Biochemistry and Physiology of Visual Pigments.* Langer, H., ed. New York: Springer-Verlag, pp. 275-282.

Connolly, K., and Cook, R. (1973): Rejection responses by female *Drosophila melanogaster:* Their ontogeny, causality and effects upon the behaviour of the courting male. *Behaviour* 44:142-166.

Constantine-Paton, M., and Law, M. I. (1978): Eye-specific termination bands in tecta of three-eyed frogs. *Science* 202:639-641.

Conway, K., Feiock, K., and Hunt, R. K. (1980): Polyclones and patterns in growing *Xenopus* eye. In *Current Topics in Developmental Biology,* Vol. 15. *Neural Development, Part 1. Emergence of Specificity in Neural Histogenesis.* Hunt, R. K., ed. New York: Academic Press, pp. 217-317.

Cook, R. (1978): The reproductive behaviour of gynandromorphic *Drosophila melanogaster. Z. Naturforsch. C* 33:744-754.

Cook, R. (1980): The extent of visual control in the courtship tracking of *Drosophila melanogaster. Biol. Cybernet.* 37:41-51.

Cook, R., and Cook, A. (1975): The attractiveness to males of female *Drosophila melanogaster:* Effects of mating, age, and diet. *Animal Behav.* 23:521-526.

Cooper, M. L., and Blasdel, G. G. (1979): Cortical topography in Siamese cats. *Soc. Neurosci. Abstr.* 5:780.

Cooper, M. L., and Pettigrew, J. D. (1979): The retinothalamic pathways in siamese cats. *J. Comp. Neurol.* 187:313-348.

Cosens, D. (1971): Blindness in a *Drosophila* mutant. *J. Insect Physiol.* 17:285-302.

Cosens, D., and Manning, A. (1969): Abnormal electroretinogram from a *Drosophila* mutant. *Nature* 224:285-287.

Cosens, D., and Perry, M. M. (1972): The fine structure of the eye of a visual mutant, A-type, of *Drosophila melanogaster. J. Insect. Physiol.* 18:1773-1786.

Costall, B., and Naylor, R. J. (1975): Cholinergic modification of

abnormal involuntary movements induced in the guinea-pig by intra-
striatal dopamine. *J. Pharm. Pharmacol.* 27:273-275.

Creel, D., Garber, S. R., King, R. A., and Witkop, C. J., Jr. (1980):
Auditory brainstem anomalies in human albinos. *Science* 209:
1253-1255.

Creel, D., O'Donnell, F. G., Jr., and Witkop, C. J., Jr. (1978): Visual
system anomalies in human ocular albinos. *Science* 201:931-933.

Crick, F. H. C., and Lawrence, P. A. (1975): Compartments and
polyclones in insect development. *Science* 189:340-347.

Cullen, M. J., and Kaiserman-Abramof, I. R. (1976): Cytological
organization of the dorsal lateral geniculate nuclei in mutant anoph-
thalmic and postnatally enucleated mice. *J. Neurocytol.* 5:407-424.

Culotti, J. G., von Ehrenstein, G., Culotti, M. R., and Russell, R. L.
(1981): A second class of acetylcholinesterase-deficient mutants of the
nematode *Caenorhabditis elegans. Genetics* 97:281-305.

Dagan, D., Kaplan, W. D., and Ikeda, K. (1975): Analysis of single
gene sex-linked behavioral mutants in *Drosophila melanogaster. Adv.
Behav. Biol.* 15:322-340.

Daly, H. V., and Sokoloff, A. (1965): *Labialpedia,* a sex-linked
mutant in *Tribolium confusum* Duval (Coleoptera: Tenebrionidae).
*J. Morphol.* 117:251-269.

Davis, R. L., and Kiger, J. A., Jr. (1981): The *dunce* mutants of
*Drosophila melanogaster:* Mutants defective in the cyclic AMP phosphodiesterase system. *J. Cell Biol.* 90:101-107.

Deak, I. I. (1976): Demonstration of sensory neurones in the ectopic
cuticle of *spineless-aristapedia,* a homeotic mutant of *Drosophila.
Nature* 260:252-254.

Deak, I. I. (1977): Mutations of *Drosophila melanogaster* that affect
muscles. *J. Embryol. Exp. Morphol.* 40:35-63.

DeFranco, A. L., and Koshland, D. G., Jr. (1980): Multiple methylation in processing of sensory signals during bacterial chemotaxis.
*Proc. Nat. Acad. Sci.* 77:2429-2433.

Deland, M. C., and Pak, W. L. (1973): Reversibly temperature-sensitive phototransduction mutant of *Drosophila melanogaster. Nature
New Biol.* 244:184-186.

Del Castillo, J., DeMello, W. C., and Morales, T. A. (1963): The
physiological role of acetylcholine in the neuromuscular system of
*Ascaris lumbricoides. Arch. Int. Physiol.* 71:741-757.

Derer, P., Caviness, V. S., Jr., and Sidman, R. L. (1977): Early cortical histogenesis in the primary olfactory cortex of the mouse. *Brain Res.* 123:27–40.

Dethier, V. G. (1969): Feeding behavior of the blowfly. *Adv. Study Behav.* 2:111–266.

Dewhurst, S. A., Croker, S. G., Ikeda, K., and McCaman, R. E. (1972): Metabolism of biogenic amines in *Drosophila* nervous tissue. *Comp. Biochem. Physiol. [B]* 43:975–981.

Dieckmann, C., and Brody, S. (1980): Circadian rhythms in *Neurospora crassa:* Oligomycin-resistant mutations affect periodicity. *Science* 207:896–898.

Dienstman, S., and Holtzer, H. (1975): Myogenesis: A cell lineage interpretation. In *Cell Cycle and Cell Differentian.* Reinert, J., and Holtzer, H., eds. New York: Springer-Verlag, pp. 1–25.

Dietrich, U., and Campos-Ortega, J. A. (1980): The effect of temperature on shibire[ts] cell clones in the compound eye of *Drosophila melanogaster. Wilhelm Roux' Arch. Entwiklungsmech. Org.* 188: 55–63.

Diliberto, E. J., Jr., and Axelrod, J. (1976): Regional and subcellular distribution of protein carboxymethylase in brain and other tissues. *J. Neurochem.* 26:1159–1165.

Dodge, F. A., Jr., Knight, B. W., and Toyoda, J. (1968): Voltage noise in *Limulus* visual cells. *Science* 160:88–90.

Dräger, U. C. (1974): Autoradiography of tritiated proline and fucose transported transneuronally from the eye to the visual cortex in pigmented and albino mice. *Brain Res.* 82:284–292.

Dräger, U. C. (1975): Receptive fields of single cells and topography in mouse visual cortex. *J. Comp. Neurol.* 160:269–290.

Dräger, U. C. (1976): Reeler mutant mice: Physiology in primary visual cortex. *Exp. Brain Res. Suppl.* 1:274–276.

Dräger, U. C. (1977): Abnormal neural development in mammals. In *Function and Formation of Neural Systems.* Stent, G. S., ed. Berlin: Dahlem Konferenzen, pp. 111–138.

Dräger, U. C. (1978): Observations on monocular deprivation in mice. *J. Neurophysiol.* 41:28–42.

Dräger, U. C., and Hubel, D. H. (1975a): Physiology of visual cells in mouse superior colliculus and correlation with somatosensory and auditory input. *Nature* 253:203–204.

Dräger, U. C., and Hubel, D. H. (1975b): Responses to visual stimulation and relationship between visual, auditory, and somatosensory inputs in mouse superior colliculus. *J. Neurophysiol.* 38:690–713.

Dräger, U. C., and Hubel, D. H. (1976): Topography of visual and somatosensory projections to mouse superior colliculus. *J. Neurophysiol.* 39:91–101.

Dräger, U. C., and Hubel, D. H. (1978): Studies of visual function and its decay in mice with hereditary retinal degeneration. *J. Comp. Neurol.* 180:85–114.

Dräger, U. C., and Olsen, J. F. (1980): Origins of crossed and uncrossed retinal projections in pigmented and albino mice. *J. Comp. Neurol.* 191:383–412.

Driskell, W. J., Weber, B. H., and Roberts, E. (1978): Purification of choline acetyltransferase from *Drosophila melanogaster. J. Neurochem.* 30:1135–1141.

Dudai, Y. (1977a): Molecular states of acetylcholinesterase from *Drosophila melanogaster. Drosophila Info. Serv.* 52:65–66.

Dudai, Y. (1977b): Properties of learning and memory in *Drosophila melanogaster. J. Comp. Physiol.* 114:69–89.

Dudai, Y. (1978): Properties of an $\alpha$-bungarotoxin-binding cholinergic nicotinic receptor from *Drosophila melanogaster. Biochem. Biophys. Acta* 539:505–517.

Dudai, Y. (1979): Behavioral plasticity in a *Drosophila* mutant, *dunce*[DB276]. *J. Comp. Physiol.* 130:271–276.

Dudai, Y., and Amsterdam, A. (1977): Nicotinic receptors in the brain of *Drosophila melanogaster* demonstrated by autoradiography with [$^{125}$I]-$\alpha$-bungarotoxin. *Brain Res.* 130:551–555.

Dudai, Y., and Bicker, G. (1978): Comparison of visual and olfactory learning in *Drosophila. Naturwissenshaften* 65:494–495.

Dudai, Y., Jan, Y. N., Byers, D., Quinn, W. G., and Benzer, S. (1976): *dunce,* a mutant of *Drosophila* deficient in learning. *Proc. Nat. Acad. Sci.* 73:1684–1688.

Duerr, J. S., and Quinn, W. G. (1981): Three mutations which block learning also affect habituation and sensitization in *Drosophila. Proc. Nat. Acad. Sci.* (in press).

Dunlap, K. (1977): Localization of calcium channels in *Paramecium caudatum. J. Physiol.* 271:119–133.

Dusenberry, D. B. (1975): The avoidance of D-tryptophan by the nematode *Caenorhabditis elegans. J. Exp. Zool.* 193:413–418.

Dusenberry, D. B. (1980): Response of the nematode *Caenorhabditis elegans* to controlled chemical stimulation. *J. Comp. Physiol.* 136: 327–331.

Dusenberry, D. B., and Barr, J. (1980): Thermal limits and chemotaxis in mutants of the nematode *Caenorhabditis elegans. J. Comp. Physiol.* 137:353–356.

Dusenberry, D. B., Sheridan, R. E., and Russell, R. L. (1975): Chemotaxis-defective mutants of the nematode *Caenorhabditis elegans. Genetics* 80:298–309.

Dvorak, D. R., Bishop, L. G., and Eckert, H. E. (1975): On the identification of movement detectors in the fly optic lobe. *J. Comp. Physiol.* 100:5–23.

Eckert, H. E., Bishop, L. G., and Dvorak, D. R. (1976): Spectral sensitivities of identified receptor cells in the blowfly *Calliphora. Naturwissenschaften* 63:47–48.

Eckert, R. (1972): Bioelectric control of ciliary activity. Locomotion in the ciliated protozoa is regulated by membrane-limited calcium fluxes. *Science* 176:473–481.

Edds, M. V., Jr., Gaze, R. M., Schneider, G. E., and Irwin, L. N. (1979): Specificity and plasticity of retinotectal connections. *Neurosci. Res. Prog. Bull.* 17:243–375.

Edgar, R. S. (1980): The genetics of development in the nematode *Caenorhabditis elegans.* In *The Molecular Genetics of Development.* Leighton, T., and Loomis, W. F., eds. New York: Academic Press, pp. 213–235.

Ede, D. A. (1956): Studies on the effects of some genetic lethal factors on the embryonic development of *Drosophila melanogaster.* IV. An analysis of the mutant X-20. *Wilhelm Roux' Arch. Entwicklungs. Org.* 149:101–114.

Ehrman, L. (1978): Sexual behavior. In *The Genetics and Biology of Drosophila,* Vol. 2b. Ashburner, M., and Wright, T. R. F., eds. London: Academic Press, pp. 127–180.

Eichenbaum, D. M., and Goldsmith, T. H. (1968): Properties of intact photoreceptor cells lacking synapses. *J. Exp. Zool.* 169:15–31.

Eicher, E. M. (1970): X-autosome translocations in the mouse: Total inactivation versus partial inactivation of the X chromosome. *Adv. Genet.* 15:175–259.

Elsner, N. (1973): The central nervous control of courtship in the grasshopper *Gomphocerippus rufus* L. (Orthoptera: Acrididae). In *Neurobiology of Invertebrates: Mechanisms of Rhythm Regulation.* Salanki, J., ed. Budapest: Akademiai Kiadó, pp. 261–287.

Emmons, S. W., Klass, M. R., and Hirsh, D. (1979): Analysis of the constancy of DNA sequences during development and evolution of the nematode *Caenorhabditis elegans*. *Proc. Nat. Acad. Sci.* 76: 1333-1337.

Englert, D. C., and Bell, A. E. (1963): *"Antennapedia:"* an unusual antennal mutation in Tribolium castaneum. *Ann. Entomol. Soc. Am.* 56:123-124.

Epp, L. G. (1972): Development of pigmentation in the eyeless mutant of the Mexican axolotl, Ambystoma mexicanum, Shaw. *J. Exp. Zool.* 181:169-180.

Epp, L. G. (1978): A review of the eyeless mutant in the Mexican axolotl. *Am. Zool.* 18:267-272.

Epstein, H. F., and Thomson, J. N. (1974): Temperature-sensitive mutation affecting myofilament assembly in *Caenorhabditis elegans*. *Nature* 250:579-580.

Ernst, A. M. (1967): Mode of action of apomorphine and dexamphetamine on gnawing compulsion in rats. *Psychopharmacologia* 10: 316-323.

Ewing, A. W. (1964): The influence of wing area on the courtship behaviour of *Drosophila melanogaster*. *Animal Behav.* 12:316-320.

Ewing, A. W., and Manning, A. (1967): The evolution and genetics of insect behaviour. *Ann. Rev. Entomol.* 12:471-494.

Falk, R., and Atidia, J. (1975): Mutation affecting taste perception in *Drosophila melanogaster*. *Nature* 254:325-326.

Fambrough, D. M. (1979): Control of acetylcholine receptors in skeletal muscle. *Physiol. Rev.* 59:165-227.

Feder, N. (1976): Solitary cells and enzyme exchange in tetraparental mice. *Nature* 263:67-69.

Feldman, J. F., and Atkinson, C. A. (1978): Genetic and physiological characteristics of a slow-growing circadian clock mutant of *Neurospora crassa*. *Genetics* 88:255-265.

Feldman, J. F., and Hoyle, M. N. (1973): Isolation of circadian clock mutants of *Neurospora crassa*. *Genetics* 75:605-613.

Felix, J. S., and DeMars, R. (1969): Purine requirement of cells cultured from humans affected with Lesch-Nyhan syndrome (hypoxanthine-guanine phosphoribosyltransferase deficiency). *Proc. Nat. Acad. Sci.* 62:536-543.

Ferrus, A., and Kankel, D. R. (1981): Cell lineage relationships in

*Drosophila melanogaster:* The relationships of cuticular to internal tissues. *Dev. Biol.* 84:485–504.

Fingerman, M. (1952): The role of the eye-pigments of Drosophila melanogaster in photic orientation. *J. Exp. Zool.* 120:131–164.

Finucci, J. M., Guthrie, J. T., Childs, A. L., Abbey, H., and Childs, B. (1976): The genetics of specific reading disability. *Ann. Human Genet.* 40:1–23.

Fischbach, K. F., and Heisenberg, M. (1981): Structural brain mutant of *Drosophila melanogaster* with reduced cell number in the medulla cortex and with normal optomotor yaw response. *Proc. Nat. Acad. Sci.* 78:1105–1109.

Flanagan, J. R. (1976): A computer program automating construction of fate maps of *Drosophila. Dev. Biol.* 53:142–146.

Flanagan, J. R. (1977): A method for fate mapping the foci of lethal and behavioral mutants in *Drosophila melanogaster. Genetics* 85: 587–607.

Flockhart, I. R., and Casida, J. E. (1972): Relationship of the acylation of membrane esterases and protein to the teratogenic action of organophosphorous insecticides and serine in developing hen eggs. *Biochem. Pharmacol.* 21:2591–2603.

Foster, M. (1965): Mammalian pigment genetics. *Adv. Genet.* 13: 311–339.

Fowler, G. L. (1973): Some aspects of the reproductive biology of *Drosophila:* Sperm transfer, sperm storage, and sperm utilization. *Adv. Genet.* 17:293–360.

Fredrickson, D. S., Gotto, A. M., and Levy, R. I. (1972): Familial lipoprotein deficiency. In *The Metabolic Basis of Inherited Disease.* Stanbury, J. B., Wyngaarden, J. B., and Fredrickson, D. S., eds. New York: McGraw-Hill, pp. 493–530.

Freeman, J. A. (1977): Possible regulatory function of acetylcholine receptor in maintenance of retinotectal synapses. *Nature* 269:218–222.

Fuchs, J. L., and Moore, R. Y. (1980): Development of circadian rhythmicity and light responsiveness in the rat suprachiasmatic nucleus: A study using the 2-deoxy [1-$^{14}$C] glucose method. *Proc. Nat. Acad. Sci.* 77:1204–1208.

Fuortes, M. G. F., and Yeandle, S. (1964): Probability of occurrence of discrete potential waves in the eye of *Limulus. J. Gen. Physiol.* 47: 443–463.

García-Bellido, A. (1979): Genetic analysis of the achaete-scute system of *Drosophila melanogaster. Genetics* 91:491–520.

García-Bellido, A., Ripoll, P., and Morata, G. (1973): Developmental compartmentalization of the wing disc of *Drosophila*. *Nature New Biol.* 245:251–253.

García-Bellido, A., and Santamaria, P. (1978): Developmental analysis of the achaete-scute system of *Drosophila melanogaster*. *Genetics* 88:469–486.

Gardner, R. L. (1978): The relationship between cell lineage and differentiation in the early mouse embryo. In *Genetic Mosaics and Cell Differentiation*, Vol. 9. *Results and Problems in Cell Differentiation*. Gehring, W. J., ed. Berlin: Springer-Verlag, pp. 205–241.

Gardner, G. F. and Feldman, J. F. (1980): The *frq* locus in *Neurospora crassa:* a key element in circadian clock organization. *Genetics* 96:877–886.

Gateff, E. (1978): Malignant neoplasms of genetic origin in *Drosophila melanogaster*. *Science* 200:1448–1459.

Gateff, E., and Schneiderman, H. A. (1974): Developmental capacities of benign and malignant neoplasms of *Drosophila*. *Wilhelm Roux' Arch. Entwicklungsmech. Org.* 176:23–65.

Gaze, R. M., Feldman, J. M., Cooke, J., and Chung, S.-H. (1979): The orientation of the visuotectal map in *Xenopus:* Developmental aspects. *J. Embryol. Exp. Morphol.* 53:39–66.

Geer, G. W., and Green, M. M. (1962): Genotype, phenotype and mating behavior of *Drosophila melanogaster*. *Am. Nat.* 96:175–181.

Gerschenfeld, H. M. (1973): Chemical transmission in invertebrate central nervous systems and neuromuscular junctions. *Physiol. Rev.* 53:1–119.

Gershon, E. S., and Jonas, W. Z. (1975): Erythrocyte soluble catechol-O-methyl transferase activity in primary affective disorder. A clinical and genetic study. *Arch. Gen. Psychiatry* 32:1351–1356.

Gershon, E. S., Targum, S. D., Kessler, L. R., Mazure, C. M., and Bunney, W. E., Jr. (1977): Genetic studies and biologic strategies in the affective disorders. *Prog. Med. Genet.* 2:101–164.

Ghysen, A. (1978): Sensory neurones recognise defined pathways in *Drosophila* central nervous system. *Nature* 274:869–872.

Ghysen, A. (1980): The projection of sensory neurons in the central nervous system of *Drosophila:* Choice of appropriate pathway. *Dev. Biol.* 78:521–541.

Ghysen, A., and Deak, I. I. (1978): Experimental analysis of sensory nerve pathways in *Drosophila*. *Wilhelm Roux' Arch. Dev. Biol.* 184:273–284.

Ghysen, A., and Janson, R. (1980): Sensory pathways in *Drosophila* central nervous system. In *Development and Neurobiology of Drosophila*. Siddiqi, O., et al., eds. New York: Plenum Press, pp. 247-265.

Gitschier, J., Strichartz, G. R., and Hall, L. M. (1980): Saxitoxin binding to sodium channels in head extracts from wild-type and tetrodotoxin-sensitive strains of *Drosophila melanogaster*. *Biochim. Biophys. Acta* 595:291-303.

Goldowitz, D., and Mullen, R. J. (1980): Weaver mutant granule defect expressed in chimeric mice. *Soc. Neurosci. Abstr.* 6:743.

Goodman, C. S. (1978): Isogenic grasshoppers: Genetic variability in the morphology of identified neurons. *J. Comp. Neurol.* 182:681-706.

Goodwin, F. K., and Post, R. M. (1975): Studies of amine metabolites in affective illness and in schizophrenia: a comparative analysis. In *Biology of the Major Psychoses: A Comparative Analysis*. Freedman, D. X., ed. New York: Raven Press, pp. 299-332.

Gorski, R. A., Gordon, J. H., Shryne, J. E., and Southam, A. M. (1978): Evidence for a morphological sex difference within the medial preoptic area of the rat brain. *Brain Res.* 148:333-346.

Gottesman, I. I. (1978): Schizophrenia and genetics: Where are we? Are you sure? In *The Nature of Schizophrenia: New Approaches to Research and Treatment*. Wynne, L. C., Cromwell, R. L., and Matthysse, S., eds. New York: John Wiley, pp. 59-69.

Götz, K. G. (1968): Flight control in *Drosophila* by visual perception of motion. *Kybernetik* 4:199-208.

Götz, K. G. (1970): Fractionation of *Drosophila* populations according to optomotor traits. *J. Exp. Biol.* 52:419-436.

Götz, K. G. (1972): Processing of cues from the moving environment in the Drosophila navigation system. In *Information Processing in the Visual Systems of Arthropods*. Wehner, R., ed. Berlin: Springer-Verlag, pp. 255-263.

Götz, K. G. (1980): Visual guidance in *Drosophila*. In *Developmental Neurobiology of Drosophila*. Siddiqi, O., et al., eds. New York: Plenum Press, pp. 391-407.

Götz, K. G. and Wenking, H. (1973): Visual control of locomotion in the walking fruitfly *Drosophila*. *J. Comp. Physiol.* 85:235-266.

Gouras, P., Carr, R. E., and Gunkel, R. D. (1971): Retinitis pigmentosa in abetalipoproteinemia: Effects of vitamin A. *Invest. Ophthalmol.* 10:784-793.

Goy, M. F., Springer, M. S., and Adler, J. (1977): Sensory transduction in *Escherichia coli:* Role of a protein methylation reaction in sensory adaptation. *Proc. Nat. Acad. Sci.* 74:4964–4968.

Goy, M. F., Springer, M. S., and Adler, J. (1978): Failure of sensory adaptation in bacterial mutants that are defective in a protein methylation reaction. *Cell* 15:1231–1240.

Green, M. C., and Sidman, R. L. (1962): Tottering—a neuromuscular mutation in the mouse. And its linkage with oligosyndactylism. *J. Hered.* 53:233–237.

Green, M. M. (1980): Transposable elements in *Drosophila* and other diptera. *Ann. Rev. Genet.* 14:109–120.

Green, S. H. (1981): The segment-specific organization of leg motoneurones in *Drosophila* is transformed in *bithorax* mutants. *Nature* 292:152–154.

Green, S. H. (1980): Innervation of antennae and legs in *Drosophila:* Wild-type and homeotic mutants. *Soc. Neurosci. Abstr.* 6:677.

Greengard, P. (1976): Possible role for cyclic nucleotides and phosphorylated membrane proteins in postsynaptic actions of neurotransmitters. *Nature* 260:101–108.

Greenspan, R. J. (1979): Genetic manipulation of neurotransmitter function in *Drosophila melanogaster* with mutants of choline acetyltransferase and acetylcholinesterase. Ph. D. Thesis, Brandeis University, Waltham, MA.

Greenspan, R. J. (1980): Mutations of choline acetyltransferase and associated neural defects in *Drosophila melanogaster. J. Comp. Physiol.* 137:83–92.

Greenspan, R. J., Finn, J. A., Jr., and Hall, J. C. (1978): Alterations in the nervous system of Drosophila melanogaster in mutants of acetylcholinesterase and choline acetyltransferase. In *Society for Neuroscience Abstracts,* Vol. IV. (Eighth Annual Meeting of the Society for Neuroscience, St. Louis, MO: Nov. 5–9, 1978), p. 194.

Greenspan, R. J., Finn, J. A., Jr., and Hall, J. C. (1980): Acetylcholinesterase mutants in *Drosophila* and their effects on the structure and function of the central nervous system. *J. Comp. Neurol.* 189:741–774.

Greenwald, I. G., and Horvitz, H. R. (1980): A behavioral mutant of *Caenorhabditis elegans* that defines a gene with a wild-type null phenotype. *Genetics* 96:147–164.

Grigliatti, T., Hall, L., Rosenbluth, R., and Suzuki, D. T. (1973): Temperature-sensitive mutations in *Drosophila melanogaster.* XIV. A selection of immobile adults. *Mol. Gen. Genet.* 120:107–114.

Grossfield, J. (1975): Behavioral mutants of Drosophila. In *Handbook of Genetics,* Vol. 3. *Invertebrates of Genetic Interest* King, R. C., ed. New York: Plenum Press, pp. 679-702.

Gruberg, E. R., and Greenhouse, G. A. (1973): The relationship of acetylcholinesterase activity to optic fiber projections in the tiger salamander *Ambystoma tigrinum. J. Morphol.* 141:147-156.

Gruberg, E. R., and Harris, W. A. (1981): The serotonergic somatosensory projection to the tectum of normal and eyeless salamanders. *J. Morphol.* (in press).

Gruberg, E. R., and Solish, S. P. (1978): The relationship of a monoamine fiber system to a somatosensory tectal projection in the salamander *Ambystoma tigrinum. J. Morphol.* 157:137-150.

Guillery, R. W. (1974): Visual pathways in albinos. *Sci. Am.* 230(5): 44-54.

Guillery, R. W., and Kaas, J. H. (1971): A study of normal and congenitally abnormal retinogeniculate projections in cats. *J. Comp. Neurol.* 143:73-100.

Guillery, R. W., Scott, G. L., Cattanach, B. M., and Deol, M. S. (1973): Genetic mechanisms determining the central visual pathways of mice. *Science* 179:1014-1016.

Hall, J. C. (1977): Portions of the central nervous system controlling reproductive behavior in *Drosophila melanogaster. Behav. Genet.* 7: 291-312.

Hall, J. C. (1978a): Behavioral analysis in *Drosophila* mosaics. In *Genetic Mosaics and Cell Differentiation.* Gehring, W. J., ed. Berlin: Springer-Verlag, pp. 259-305.

Hall, J. C. (1978b): Courtship among males due to a male-sterile mutation in *Drosophila melanogaster. Behav. Genet.* 8:125-141.

Hall, J. C. (1979): Control of male reproductive behavior by the central nervous system of Drosophila: Dissection of a courtship pathway by genetic mosaics. *Genetics* 92:437-457.

Hall, J. C., Alahiotis, S. N., Strumpf, D. A., and White, K. (1980a): Behavioral and biochemical defects in temperature-sensitive acetylcholinesterase mutants of *Drosophila melanogaster. Genetics* 96: 939-965.

Hall, J. C., Gelbart, W. M., and Kankel, D. R. (1976): Mosaic systems. In *The Genetics and Biology of Drosophila,* Vol. 1a. Ashburner, M., and Novitski, E., eds. London: Academic Press, pp. 265-314.

Hall, J. C., and Greenspan, R. J. (1979): Genetic analysis of *Drosophila* neurobiology. *Ann. Rev. Genet.* 13:127-195.

Hall, J. C., Greenspan, R. J., and Kankel, D. R. (1979): Neural defects induced by genetic manipulation of acetylcholine metabolism in *Drosophila*. In *Society for Neuroscience Symposia*, Vol. 4. Ferendeli, J. A., ed. Bethesda, MD: Society for Neuroscience, pp. 1-42.

Hall, J. C., and Kankel, D. R. (1976): Genetics of acetylcholinesterase in *Drosophila melanogaster*. *Genetics* 83:517-535.

Hall, J. C., Siegel, R. W., Tompkins, L., and Kyriacou, C. P. (1980b): Neurogenetics of courtship in *Drosophila*. *Stadler Genet. Symp.* 12:43-82.

Hall, J. C., Tompkins, L., Kyriacou, C. P., Siegel, R. W., Schilcher, F. V., and Greenspan, R. J. (1980c): Higher behavior in *Drosophila* analyzed with mutations that disrupt the structure and function of the nervous system. In *Development and Neurobiology of Drosophila*. Siddiqi, O., et al., eds. New York: Plenum Press, pp. 425-455.

Hall, L. M. (1980): Biochemical and genetic analysis of an α -bungarotoxin-binding reception from *Drosophila melanogaster*. In *Receptors for Neurotransmitters, Hormones and Pheromones in Insects*. Satelle, D. B., Hall, L. M., and Hildebrand, J. G., eds. Amsterdam: Elsevier/North-Holland Biomedical Press, pp. 111-124.

Hall, L. M., Von Borstel, R. W., Osmond, B. C., Hoeltzli, S. D., and Hudson, T. H. (1978): Genetic variants in an acetylcholine receptor from *Drosophila melanogaster*. *FEBS Lett.* 95:243-246.

Hall, Z. W., Hildebrand, J. G., and Kravitz, E. A. (1974): *Chemistry of Synaptic Transmission*. Newton, MA: Chiron Press.

Hamburgh, M., Peterson, E., Bornstein, M. B., and Kirk, C. (1975): Capacity of feotal spinal cord obtained from dystrophic mice ($dy^2J$) to promote muscle regeneration. *Nature* 256:219-220.

Handler, A. M., and Konopka, R. J. (1978): Transplantation of a circadian oscillator in *Drosophila melanogaster*. *Physiologist* 21:50.

Handler, A. M., and Konopka, R. J. (1979): Transplantation of a circadian pacemaker in *Drosophila*. *Nature* 279:236-238.

Hansen, K. (1969): The mechanism in insect sugar reception, a biochemical investigation. In *Olfaction and Taste*, Vol. 3. Pfaffmann, C., ed. New York: Rockefeller University Press, pp. 382-391.

Hansen, K. and Strausfeld, N. J. (1980): Sexually dimorphic interneuron arrangements in the fly visual system. *Proc. Roy. Soc. B* 208: 57-71.

Hansma, H. G., and Kung, C. (1975): Studies of the cell surface of *Paramecium*. Ciliary membrane proteins and immobilization antigens. *Biochem. J.* 152:523–528.

Hardie, R. C., Franceschini, N., and McIntyre, P. D. (1979): Electrophysiological analysis of fly retina. II. Spectral and polarisation sensitivity in R7 and R8. *J. Comp. Physiol.* 133:23–40.

Harris, W. A. (1972): The maxillae of *Drosophila melanogaster* as revealed by scanning electron microscopy. *J. Morphol.* 138:451–456.

Harris, W. A. (1979): Amphibian chimeras and the nervous system. In *Society for Neuroscience Symposia*, Vol. IV. *Aspects of Developmental Neurobiology*. Ferrendelli, J. A., ed. Bethesda: Society for Neuroscience, pp. 228–257.

Harris, W. A. (1981): The effects of eliminating impulse activity on the development of the retinotectal projection in salamanders. *J. Comp. Neurol.* 194:303–317.

Harris, W. A. (1980): Physiological activity and development. *Ann. Rev. Physiol.* 43:689–710.

Harris, W. A., and Stark, W. S. (1977): Hereditary retinal degeneration in *Drosophila melanogaster*. A mutant defect associated with the phototransduction process. *J. Gen. Physiol.* 69:261–291.

Harris, W. A., Stark, W. S., and Walker, J. A. (1976): Genetic dissection of the photoreceptor system in the compound eye of *Drosophila melanogaster*. *J. Physiol.* 256:415–439.

Harrison, R. G. (1904): Experimentelle Untersuchungen über die Entwicklung der Sinnesorgane der Seitenlinie bei den Amphibien. *Arch. Mikrosk. Anat.* 63:35–149.

Harrison, R. G. (1924): Some expected results of the heteroplastic transplantation of limbs. *Proc. Nat. Acad. Sci.* 10:69–74.

Harrison, R. G. (1929): Correlation in the development and growth of the eye studied by means heteroplastic transplantation. *Wilhelm Roux' Arch. Entwiklungsmech. Org.* 120:1–55.

Harrison, R. G. (1934): Heteroplastic grafting in embryology. *Harvey Lect.* 29:116–157.

Hatten, M. E., and Sidman, R. L. (1977): Plant lectins detect age and region specific differences in cell surface carbohydrates and cell reassociation behavior of embryonic mouse cerebellar cells. *J. Supramol. Struct.* 7:267–275.

Hatten, M. E., and Sidman, R. L. (1978): Cell reassociation behavior and lectin-induced agglutination of embryonic mouse cells from different brain regions. *Exp. Cell Res.* 113:111–125.

Hatten, M. E., Trenkner, E., and Sidman, R. L. (1976): Cell migration and cell-cell interactions in primary cultures of embryonic mouse cerebellum. *Soc. Neurosci. Abstr.* 2:1023.

Hausen, K., and Strausfeld, N. J. (1980): Sexually dimorphic interneuron arrangements in the fly visual system. *Proc. Roy. Soc. B* 208: 57–71.

Hazelbauer, G. L. (1975): Maltose chemoreceptor of *Escherichia coli*. *J. Bacteriol.* 122:206–214.

Hazelbauer, G. L., and Adler, J. (1971): Role of the galactose binding protein in chemotaxis of *Escherichia coli* toward galactose. *Nature New Biol.* 230:101–104.

Hearing, V. J. (1973): Tyrosinase activity in subcellular fractions of black and albino mice. *Nature New Biol.* 245:81–83.

Hedgecock, E. M., and Russell, R. L. (1975): Normal and mutant thermotaxis in the nematode *Caenorhabditis elegans*. *Proc. Nat. Acad. Sci.* 72:4061–4065.

Hedin, P. A., Niemeyer, C. S., Gueldner, R. C., and Thompson, A. C. (1972): A gas chromatographic survey of the volatile fractions of twenty species of insects from eight orders. *J. Insect. Physiol.* 18:555–564.

Heisenberg, M. (1971): Separation of receptor and lamina potentials in the electroretinogram of normal and mutant *Drosophila*. *J. Exp. Biol.* 55:85–100.

Heisenberg, M. (1972): Comparative behavioral studies on two mutants of *Drosophila*. *J. Comp. Physiol.* 80:119–136.

Heisenberg, M. (1980): Mutants of brain structure and function: What is the significance of the mushroom bodies for behavior? In *Development and Neurobiology of Drosophila*. Siddiqi, O., et al., eds. New York: Plenum Press, pp. 373–390.

Heisenberg, M., and Bohl, K. (1979): Isolation of anatomical brain mutants of *Drosophila* by histological means. *Z. Naturforsch.* 34: 143–147.

Heisenberg, M., and Buchner, E. (1977): The role of retinula cell types in visual behavior of *Drosophila melanogaster*. *J. Comp. Physiol.* 117:127–162.

Heisenberg, M., and Götz, K. G. (1975): The use of mutations for the partial degradation of vision in *Drosophila melanogaster. J. Comp. Physiol.* 98:217-241.

Heisenberg, M., and Wolf, R. (1979): On the fine structure of yaw torque in visual orientation of *Drosophila melanogaster. J. Comp. Physiol.* 130:113-130.

Heisenberg, M., Wonneberger, R., and Wolf, R. (1978): optomotor-blind[H31]-*Drosophila* mutant of the lobula plate giant neurons. *J. Comp. Physiol. A* 124:287-296.

Hennessey, T., and Nelson, D. L. (1979): Thermosensory behavior in *Paramecium tetraurelia:* A quantitative assay and some factor that influence thermal avoidance. *J. Gen. Microbiol.* 112:337-347.

Herrup, K., and Mullen, R. J. (1979): Staggerer Chimeras: Intrinsic nature of Purkinje cell defects and implications for normal cerebellar development. *Brain Res.* 178:443-457.

Hibbard, E., and Ornberg, R. L. (1976): Restoration of vision in genetically eyeless axolotls. *Exp. Neurol.* 50:113-123.

Hildebrand, J. G., Hall, L. M., and Osmond, B. C. (1979): Distribution of binding sites for $^{125}$I-labeled $\alpha$-bungarotoxin in normal and deafferented antennal lobes of *Manduca sexta. Proc. Nat. Acad. Sci.* 76:499-503.

Hirsch, D., and Vanderslice, R. (1976): Temperature-sensitive developmental mutants of *Caenorhabditis elegans. Dev. Biol.* 49:220-235.

Hodgetts, R. B. (1975): The response of dopa decarboxylase activity to variations in gene dosage in *Drosophila:* A possible location of the structural gene. *Genetics* 79:45-54.

Hodgkin, J. A., and Brenner, S. (1977): Mutation causing transformation of sexual phenotype in the nematode *Caenorhabditis elegans. Genetics* 86:275-287.

Holliday, R., and Pugh, J. E. (1975): DNA modification mechanisms and gene activity during development. *Science* 187:226-232.

Homyk, T., Jr., Szidonya, J., and Suzuki, D. T. (1980): Behavioral mutants of *Drosophila melanogaster.* III. Isolation and mapping of mutations by direct visual observations of behavioral phenotypes. *Mol. Gen. Genet.* 177:553-565.

Horridge, G. A. (1962): Learning of leg position by the ventral nerve cord in headless insects. *Proc. Roy. Soc. B* 157:33-52.

Horridge, G. A. and Meinertzhagen, I. A. (1970): The accuracy of the patterns of connexions of the first- and second-order neurons of the visual system of *Calliphora. Proc. Roy. Soc. B* 175:69-82.

Horvitz, H. R. (1981): Neuronal cell lineages in the nematode *Caenorhabditis elegans*. In *Development of the Nervous System*. Garrod, D. R., and Feldman, J. D., eds. New York: Cambridge University Press, pp. 331-345.

Horvitz, H. R., and Sulston, J. (1981): Isolation and genetic characterization of cell lineage mutants of the nematode *Caenorhabditis elegans*. *Genetics* 96:435-454.

Hotta, Y. (1979): A biochemical analysis of visual mutations in *Drosophila melanogaster:* Changes in major eye proteins. In *Mechanisms of Cell Change*. Ebert, J. D., and Okada, T., eds. New York: John Wiley, pp. 169-182.

Hotta, Y., and Benzer, S. (1969): Abnormal electroretinograms in visual mutants of *Drosophila. Nature* 222:354-356.

Hotta, Y., and Benzer, S. (1970): Genetic dissection of the *Drosophila* nervous system by means of mosaics. *Proc. Nat. Acad. Sci.* 67: 1156-1163.

Hotta, Y., and Benzer, S. (1972): Mapping of behaviour in *Drosophila* mosaics. *Nature* 240:527-535.

Hotta, Y., and Benzer, S. (1973): Mapping of behavior in *Drosophila* mosaics. *Symp. Soc. Dev. Biol.* 31:129-167.

Hotta, Y., and Benzer, S. (1976): Courtship in *Drosophila* mosaics: Sex-specific foci for sequential action patterns. *Proc. Nat. Acad. Sci.* 73:4154-4158.

Howse, P. E. (1975): Brain structure and behavior in insects. *Ann. Rev. Entomol.* 20:359-379.

Hoy, R. R. (1974): Genetic control of acoustic behavior in crickets. *Am. Zool.* 14:1067-1080.

Hoy, R. R., Hahn, J., and Paul, R. C. (1976): Hybrid cricket auditory behavior: Evidence for genetic coupling in animal communication. *Science* 195:82-84.

Hu, K. G., Reichert, H., and Stark, W. S. (1978): Electrophysiological characterization of *Drosophila ocelli. J. Comp. Physiol.* 126: 15-24.

Hu, K. G., and Stark, W. S. (1980): The roles of *Drosophila ocelli* and compound eyes in phototaxis. *J. Comp. Physiol.* 135:85-95.

Hubel, D. H., and Wiesel, T. N. (1971): Aberrant visual projections in the Siamese cat. *J. Physiol.* 218:33-62.

Huber, F. (1965): Brain controlled behavior in orthopterans. In *The Physiology of the Insect Central Nervous System*. Beament, J. W. L.,

Treherne, J. E., and Wigglesworth, V. B., eds. London: Academic Press, pp. 233-246.

Huber, F. (1967): Central control of movements and behavior of invertebrates. In *Invertebrate Nervous Systems: Their Significance for Mammalian Neurophysiology.* Wiersma, C. A. G., ed. Chicago: University of Chicago Press, pp. 333-351.

Huber, F. (1974): Neural integration (central nervous system). In *Physiology of Insecta, 2nd Edition,* Vol. 4. Rockstein, M., ed. New York: Academic Press, pp. 3-100.

Humphrey, R. R. (1969): A recently discovered mutant "eyeless" in the Mexican axolotl. *Ambystoma mexicanum. Anat. Rec.* 163:206.

Humphrey, R. R. (1972): Genetic and experimental studies on a mutant gene ($c$) determining absence of heart action in embryos of the Mexican axolotl *Ambystoma mexicanum. Dev. Biol.* 27:365-375.

Humphrey, R. R. (1975): The axolotl, *Ambystoma mexicanum.* In *Handbook of Genetics,* Vol. 4. King, R. C., ed. New York: Plenum Press, pp. 3-17.

Humphrey, R. R., and Chung, H. M. (1977): Genetic and experimental studies on three associated mutant genes in the Mexican axolotl: st (for statis), mi (for microphthalmic) and h (for hand lethal). *J. Exp. Zool.* 202:195-202.

Hunt, R. K. (1975a): The cell cycle, cell lineage, and neuronal specificity. In *Cell Cycle and Cell Differentiation. (Results and Problems in Cell Differentiation Series,* Vol. 7.) Reinert, J., and Holtzer, H., eds. Heidelberg: Springer-Verlag, pp. 43-62.

Hunt, R. K. (1975b): Development programming for retinotectal patterns. In *Cell Patterning.* (Ciba Foundation Symposium 29, new series) New York: American Elsevier, pp. 131-159.

Hunt, R. K. (1975c): Position-dependent differentiation of neurons. In *Developmental Biology. Pattern Formation. Gene Regulation.* McMahon, D., and Fox, C. F., eds. Reading, MA: Benjamin, pp. 227-256.

Hunt, R. K., and Berman, N. (1975): Patterning of neuronal locus specificities in retinal ganglion cells after partial extirpation of the embryonic eye. *J. Comp. Neurol.* 162:43-70.

Ide, C. F. (1977): Neurophysiology of *spastic,* a behavior mutant of the Mexican axolotl: Altered vestibular projection to cerebellar auricle and area acoustico-lateralis. *J. Comp. Neurol.* 176:359-372.

Ide, C. F. (1978): Genetic dissection of cerebellar development: Mutations affecting cell position. *Am. Zool.* 18:281-287.

Ide, C. F., Miszkowski, N., Kimmel, C. B., Schabtach, E., Tompkins, R., Elbert, O., and Duda, M. (1977a): Analysis of spastic: A neurological mutant of the Mexican axolotl. *Prog. Clin. Biol. Res.* 15: 267–289.

Ide, C. F., and Tompkins, R. (1975): Development of locomotor behavior in wild type and *spastic* (sp/sp) axolotls, *Ambystoma mexicanum. J. Exp. Zool.* 194:467–478.

Ide, C. F., Tompkins, R., and Miszkowski, N. (1977b): Neuroanatomy of *Spastic,* a behavior mutant of the Mexican axolotl: Purkinje cell distribution in the adult cerebellum. *J. Comp. Neurol.* 176: 373–386.

Ikeda, K., and Kaplan, W. D. (1970a): Patterned neural activity of a mutant *Drosophila melanogaster. Proc. Nat. Acad. Sci.* 66:765–772.

Ikeda, K., and Kaplan, W. D. (1970b): Unilaterally patterned neural activity of gynandromorphs, mosaic for a neurological mutant of *Drosophila melanogaster. Proc. Nat. Acad. Sci.* 67:1480–1487.

Ikeda, K., and Kaplan, W. D. (1974): Neurophysiological genetics in *Drosophila melanogaster. Am. Zool.* 14:1055–1066.

Ikeda, K., Ozawa, S., and Hagiwara, S. (1976): Synaptic transmission reversibly conditioned by single-gene mutation in *Drosophila melanogaster. Nature* 259:489–491.

Ingham, P., and Whittle, R. (1980): Trithorax—a new homoeotic mutation of *Drosophila melanogaster* causing transformation of abdominal and thoracic imaginal segments. *Mol. Gen. Genet.* 179: 607–614.

Isono, K., and Kikuchi, T. (1974): Autosomal recessive mutation in sugar response of *Drosophila. Nature* 248:243–244.

Jacob, K. G., Willmund, R., Folkers, E., Fischbach, K. F., and Spatz, H. C. (1977): T-maze phototaxis of *Drosophila melanogaster* and several mutants in the visual systems. *J. Comp. Physiol.* 116: 209–225.

Jacobson, C.-O. (1959): The localization of the presumptive cerebral regions in the neural plate of the axolotl larva. *J. Emb. Exp. Morph.* 7:1–21.

Jacobson, M. (1978): *Developmental Neurobiology,* 2nd ed. New York: Plenum Press.

Jacobson, M., and Hirose, G. (1978): Origin of the retina from both sides of the embryonic brain: A contribution to the problem of crossing at the optic chiasma. *Science* 202:637–639.

Jallon, J.-M., and Hotta, Y. (1979): Genetic and behavioral studies of female sex appeal in *Drosophila*. *Behav. Genet.* 9:257-275.

Jan, L. Y., and Jan, Y. N. (1976a): L-glutamate as an excitatory transmitter at the *Drosophila* larval neuromuscular junction. *J. Physiol.* 262:215-236.

Jan, L. Y., and Jan, Y. N. (1976b): Properties of the larval neuromuscular junction in *Drosophila melanogaster*. *J. Physiol.* 262: 189-214.

Jan, L. Y., and Jan, Y. N. (1981): Mutations of the synapse in Drosophila. In *Development of the Nervous System*. Garrod, D. R., and Feldman, J. D., eds. Cambridge: Cambridge University Press, pp. 347-360.

Jan, Y. N., and Jan, L. Y. (1978): Genetic dissection of short-term and long-term facilitation at the *Drosophila* neuromuscular junction. *Proc. Nat. Acad. Sci.* 75:515-519.

Jan, Y. N., Jan, L. Y., and Dennis, M. J. (1977): Two mutations of synaptic transmission in *Drosophila*. *Proc. Roy. Soc. B* 198:87-108.

Janning, W. (1978): Gynandromorph fate maps in *Drosophila*. In *Genetic Mosaics and Cell Differentiation*. Gehring, W. J., ed. Berlin: Springer-Verlag, pp. 1-28.

Jennings, H. S. (1906): *Behavior of the Lower Organisms*. New York: Columbia University Press.

Jiménez, F., and Campos-Ortega, J. A. (1979): A region of the *Drosophila* genome necessary for CNS development. *Nature* 282: 310-312.

Johnson, C. D., Duckett, J. G., Culotti, J. G., Herman, R. K., Meneely, P. M., and Russell, R. L. (1981): An acetylcholinesterase-deficient *Caenorhabditis elegans*. *Genetics* 97:261-279.

Johnson, L. M., and Sidman, R. L. (1979): A reproductive endocrine profile in the diabetes (db) mutant mouse. *Biol Reproduct.* 20: 552-559.

Kaas, J. H., and Guillery, R. W. (1973): The transfer of abnormal visual field representations from the dorsal lateral geniculate nucleus to the visual cortex in Siamese cats. *Brain Res.* 59:61-95.

Kalckar, H. M. (1971): The periplasmic galactose binding protein of *Escherichia coli*. *Science* 174:557-565.

Kandel, E. R. (1976): *The Cellular Basis of Behavior: An Introduction to Behavioral Neurobiology*. San Francisco: W. H. Freeman.

Kandel, E. R. (1978): *A Cell-Biological Approach to Learning*. (Grass Lecture Monograph 1) Bethesda: Society for Neuroscience.

Kankel, D. R., and Hall, J. C. (1976): Fate mapping of nervous system and other internal tissues in genetic mosaics of *Drosophila melanogaster*. *Dev. Biol.* 48:1-24.

Kaplan, W. D. (1979): Motor activity mutants of *Drosophila*. In *Psychology Survey 2*. Connolly, K., ed. London: Allen and Unwin, pp. 90-109.

Kaplan, W. D., and Trout, W. E., III (1969): The behavior of four neurological mutants of *Drosophila*. *Genetics* 61:399-409.

Katz, B., and Miledi, R. (1968): The role of calcium in neuromuscular facilitation. *J. Physiol.* 195:481-492.

Katz, B., and Miledi, R. (1969): Tetrodotoxin-resistant electrical activity in presynaptic terminals. *J. Physiol.* 203:459-487.

Katz, B., and Thesleff, S. (1957a): The interaction between edrophonium (tensilon) and acetylcholine at the motor end-plate. *Brit. J. Pharmacol. Chemother.* 12:260-264.

Katz, B., and Thesleff, S. (1957b): A study of the "densensitization" produced by acetylcholine at the motor end-plate. *J. Physiol.* 138: 63-80.

Kauffman, S. A. (1973): Control circuits for determination and transdetermination. *Science* 181:310-318.

Kauffman, S. A. (1975): Control circuits for determination and transdetermination: interpreting positional information in a binary epigenetic code. In *Cell Patterning*. (Ciba Foundation Symposium 29.) Amsterdam: Elsevier-Excerpta Medica-North-Holland, pp. 201-221.

Kaufmann, S. A., Shymko, R. M., and Trabert, K. (1978): Control of sequential compartment formation in *Drosophila*. *Science* 199: 259-270.

Kelly, L. E. (1981): Correlation of an unusual "light-off" induced escape response with alteration in the electororetinogram off-transient of a *Drosophila* mutant. *J. Comp. Physiol.* (in press).

Kelly, L. E., and Suzuki, D. T. (1974): The effects of increased temperature on electroretingrams of temperature-sensitive paralytic mutants in *Drosophila melanogaster*. *Proc. Nat. Acad. Sci.* 71: 4906-4909.

Kennedy, C., DesRosier, M., Jehle, J. W., Reivich, F., and Sokoloff, L. (1975): Mapping of functional neural pathways by autoradiographic survey of local metabolic rate with [$^{14}$C] deoxyglucose. *Science* 187:850–853.

Kety, S. S., Rosenthal, D., Wender, P. H., Schulsinger, F., and Jacobsen, B. (1978): The biologic and adoptive families of adopted individuals who became schizophrenic: prevalence of mental illness and other characteristics. In *The Nature of Schizophrenia. New Approaches to Research and Treatment.* Wynne, L. C., Cromwell, R. L., and Matthysee, S., eds. New York: John Wiley, pp. 25–37.

Khan, S., Macnab, R. M., DeFranco, A. L., and Koshland, D. E. (1978): Inversion of a behavioral response in bacterial chemotoxis: Explanation at the molecular level. *Proc. Nat. Acad. Sci.* 75: 4150–4154.

Kidd, K. K. (1978): A genetic perspective on schizophrenia. In *The Nature of Schizophrenia: New Approaches to Research and Treatment.* Wynne, L. C., Cromwell, R. L., and Matthysse, S., eds. New York: John Wiley, pp. 70–75.

Kiger, J. A., Jr. (1977): The consequences of nullosomy for a chromosomal region affecting cyclic AMP phosphodiesterase activity in *Drosophila. Genetics* 85:623–628.

Kiger, J. A., Jr., Davis, R. L., Salz, H., Fletcher, T., and Bowling, M. (1981): Genetic analysis of cyclic nucleotide phosphodiesterases in *Drosophila melanogaster. Adv. Cyclic Nucleotide Res.* 14:273–288.

Kiger, J. A., Jr., and Golanty, E. (1977): A cytogenetic analysis of cyclic nucleotide phosphodiesterase in *Drosophila. Genetics* 85: 609–622.

Kiger, J. A., Jr., and Golanty, E. (1979): A genetically distinct form of cyclic AMP phosphodiesterase associated with chromomere 3D4 in *Drosophila melanogaster. Genetics* 91:521–535.

Kinney, D. K., and Jacobsen, B. (1978): Environmental factors in schizophrenia: new adoption study evidence. In *The Nature of Schizophrenia. New Approaches to Research and Treatment.* Wynne, L. C., Cromwell, R. L., and Matthysse, S., eds. New York: John Wiley, pp. 38–51.

Kirk, D. L. (1976): Projections of the visual field by the axons of cat retinal ganglion cells. Section 2. Decussation of optic axons in Siamese cats. Ph. D. Thesis, Australian National University, Canberra, Australia, pp. 1–17.

Kirschfeld, K. (1967): Die Projektion der optischen Umwelt auf das

Raster der Rhabdomere in Komplexauge von Musca. *Exp. Brain Res.* 3:248-270.

Kirschfeld, K. (1971): Aufnahme und Vararbeitung optischer Daten im Komplexauge von Insekter. *Naturwissenschaften* 58:201-209.

Kirschfeld, K. (1972): The visual system of *Musca:* Studies on optics, structure and function. In *Information Processing in the Visual System of Arthropods.* Wehner, R., ed. Berlin: Springer-Verlag, pp. 61-74.

Kirschfeld, K. (1973): Das neurale Superpositionsauge. *Fortschr. Zool.* 21:229-257.

Kirschfeld, K., and Lutz, B. (1974): Lateral inhibition in the compound eye of the fly, *Musca. Z. Naturforsch. [c]* 29:95-97.

Kirschner, D. A., and Sidman, R. L. (1976): X-ray diffraction study of myelin structure in immature and mutant mice. *Biochim. Biophys. Acta* 448:73-87.

Kleene, S. J., Hobson, A. C., and Adler, J. (1979): Attractants and repellents influence methylation and demethylation of methly-accepting chemotaxis proteins in an extract of *Escherichia coli. Proc. Nat. Acad. Sci.* 76:6309-6313.

Klemm, N. (1976): Histochemistry of putative transmitter substances in the insect brain. *Prog. Neurobiol.* 7:99-169.

Koenig, J. H., and Ikeda, K. (1980): Flight pattern induced by temperature in a single-gene mutant of *Drosophila melanogaster. J. Neurobiol.* 11:509-517.

Kondoh, H., Ball, C. B., and Adler, J. (1979): Identification of a methyl-accepting chemotaxis protein for the ribose and galactose chemoreceptors of *Escherichia coli. Proc. Nat. Acad. Sci.* 76: 260-264.

Konishi, M., and Nottebohm, F. (1969): Experimental studies in the ontogeny of avian vocalizations. In *Bird Vocalizations. Their Relation to Current Problems in Biology and Psychology.* Hinde, R. A., ed. Cambridge: Cambridge University Press, pp. 29-48.

Konopka, R. J. (1972): Circadian clock mutants of *Drosophila melanogaster.* Ph. D. Thesis, California Institute of Technology, Pasadena, CA.

Konopka, R. J. (1976): Genetic dissection of complex behavior. In *The Molecular Basis of Circadian Rhythms.* Hastings, J. W., and Schweiger, H.-G., eds. Berlin: Dahlem Conference, 1975, pp. 327-337.

Konopka, R. J. (1979): Genetic dissection of the *Drosophila* circadian system. *Fed. Proc.* 38:2602–2605.

Konopka, R. J. (1981): Genetics and development of circadian rhythms in invertebrates. In *Handbook of Behavioral Neurobiology,* Vol. 5. Aschoff, J., ed. New York: Plenum, pp. 173–181.

Konopka, R. J., and Benzer, S. (1971): Clock mutants of *Drosophila melanogaster. Proc. Nat. Acad. Sci.* 68:2112–2116.

Konopka, R. J., Pittendrigh, C. S., and Orr, D. (1981a): Reciprocal behavior, impaired homeostasis, and altered photosensitivity of *Drosophila* clock mutants, *Am. J. Physiol.* (in press).

Konopka, R. J., and Wells, S. (1980): *Drosophila* clock mutations affect the morphology of a brain neurosecretory cell group. *J. Neurobiol.* 11:411–415.

Konopka, R. J., and Wells, S. (1981): A circadian rhythm of cell division in *Drosophila* cell culture. *Science* (in press).

Konopka, R. J., Wells, S., and Lee, T. (1981b): Mosaic analysis of a Drosophila clock mutant. *Dev. Genet.* (in press).

Kort, E. N., Goy, M. F., Larsen, S. H., and Adler, J. (1975): Methylation of a membrane protein involved in bacterial chemotaxis. *Proc. Nat. Acad. Sci.* 72:3939–3943.

Koshland, D. E., Jr. (1979): A model regulatory system-bacterial chemotaxis. *Physiol. Rev.* 59:811–862.

Koshland, D. E., Jr. (1980): Bacterial chemotaxis in relation to neurobiology. *Ann. Rev. Neurosci.* 3:43–75.

Kung, C. (1976): Membrane control of ciliary motions and its genetic modification. In *Cell Motility. Book C.* Goldman, R., Pollard, T., and Rosenbaum, J., eds. New York: Cold Spring Laboratory, pp. 941–948.

Kung, C. (1979): Biology and genetics of *Paramecium* behavior. In *Neurogenetics: Genetic Approaches to the Nervous System.* Breakefield, X. O., editor-in-chief. New York: Elsevier North-Holland, pp. 1–26.

Kung, C., Chang, S.-Y., Satow, Y., Van Houten, J., and Hansma, H. (1975): Genetic dissection of behavior in *Paramecium. Science* 188: 898–904.

Kyriacou, C. P., and Hall, J. C. (1980): Circadian rhythm mutations in *Drosophila melanogaster* affect short-term fluctuations in the male's courtship song. *Proc. Nat. Acad. Sci.* 77:6729–6733.

Lagerspatz, K. M. J., and Lagerspatz, K. Y. H. (1974): Genetic determination of aggressive behaviour. In *The Genetics of Behaviour.* van Abeelen, J. H. F., ed. Amsterdam: Elsevier/North-Holland Biomedical Press, pp. 321-346.

Lake, C. R., and Ziegler, M. G. (1977): Lesch-Nyhan syndrome: Low dopamine-$\beta$-hydroxylase activity and diminished sympathetic response to stress and posture. *Science* 196:905-906.

Landauer, W. (1975): Cholinomimetic teratogens: Studies with chicken embryos. *Teratology* 12:125-146.

Landis, D. M. D., and Reese, T. S. (1977): Structure of the Purkinje cell membrane in staggerer and weaver mutant mice. *J. Comp. Neurol.* 171:247-260.

Landis, D. M. D., and Sidman, R. L. (1978): Electron microscopic analysis of postnatal histogenesis in the cerebellar cortex of staggerer mutant mice. *J. Comp. Neurol.* 179:831-864.

Lanna, T. M., and Franco, M. G. (1961): Modalità di ricombinazione di alcuni mutanti di *Musca domestica. Genet. Agraria* 14:297-306.

Larsen, S. H., Reader, R. W., Kort, E. N., Tso, W.-W., and Adler, J. (1974): Change in direction of flagellar rotation is the basis of the chemotactic response in *Escherichia coli. Nature* 249:74-77.

LaVail, J. H., Nixon, R. A., and Sidman, R. L. (1978): Genetic control of retinal ganglion cell projections. *J. Comp. Neurol.* 182: 399-421.

LaVail, M. M., and Mullen, R. J. (1976): Role of the pigment epithelium in inherited retinal degeneration analyzed with experimental mouse chimeras. *Exp. Eye Res.* 23:227-245.

LaVail, M. M., Sidman, R. L., and Gerhardt, C. O. (1975): Congenic strains of RCS rats with inherited retinal dystrophy. *J. Hered.* 66: 242-244.

Law, P. K. (1977): "Myotropic" influence on motoneurons of normal and dystrophic mice in parabiosis. *Exp. Neurol.* 54:444-452.

Law, P. Y., Harris, R. A., Loh, H. H., and Way, E. L. (1978): Evidence for the involvement of cerebroside sulfate in opiate receptor binding: Studies with Azure A and jimpy mutant mice. *J. Pharmacol. Exp. Ther.* 207:458-468.

Lawrence, P. A. (1978): Compartments and the insect nervous system. *Zoon* 6:157-160.

Lawrence, P. A., and Green, S. M. (1979): Cell lineage in the developing retina of *Drosophila. Dev. Biol.* 71:142-152.

Lemanski, L. F, Paulson, D. J., and Hill, C. S. (1979): Normal anterior endoderm corrects the heart defect in cardiac mutant sala-manders (*Ambystoma mexicanum*). *Science* 204:860–862.

Leonard, J. P., and Salpeter, M. M. (1979): Agonist-induced myopathy at the neuromuscular junction is mediated by calcium. *J. Cell Biol.* 82:811–819.

Levine, J. D. (1974): Giant neuron input in mutant and wild type *Drosophila*. *J. Comp. Physiol.* 93:265–285.

Levinthal, C. (1974): Neural development in isogenic organisms—a summary report. *Am. Zool.* 14:1051–1054.

Levinthal, F., Macagno, E., and Levinthal, C. (1975): Anatomy and development of identified cells in isogenic organisms. *Cold Spring Harbor Symp. Quant. Biol.* 40:321–331.

Lewis, E. B. (1963): Genes and developmental pathways. *Am. Zool.* 3:33–56.

Lewis, E. B. (1978): A gene complex controlling segmentation in *Drosophila*. *Nature* 276:565–570.

Lewis, E. B. (1980): Report of *bx⁷:* Bithorax-7. *Drosophila Inf. Serv.* 55:207–208.

Lewis, J. A., and Hodgkin, J. A. (1977): Specific neuroanatomical changes in chemosensory mutants of the nematode *Caenorhabditis elegans*. *J. Comp. Neurol.* 172:489–510.

Lewis, J. A., Wu, C.-H., Levine, J. H., and Berg, H. (1980): Levamisole-resistant mutants of the nematode *Caenorhabditis elegans* appear to lack pharmacological acetylcholine receptors. *Neuroscience* 5:967–989.

Lewis, W. H. (1907): Transplantation of the lips of the blastopore in *Rana palustris*. *Am. J. Anat.* 7:137–143.

Liebman, P. A., and Pugh, E. N., Jr. (1979): The control of phosphodiesterase in rod disk membranes: Kinetics, possible mechanisms and significance for vision. *Vision Res.* 19:375–380.

Lindsley, D. L., and Grell, E. H. (1968): *Genetic Variations of Drosophila melanogaster*. Washington, D.C.: Carnegie Institution Publication No. 627.

Lindsley, D. L., Sandler, S., Baker, B. S., Carpenter, A. T. C., Denell, R. E., Hall, J. C., Jacobs, P. A., Miklos, G. L. G., Davis, B. K., Gethmann, R. C., Hardy, R. W., Hessler, A., Miller, S. M., Nozawa, H., Parry, D. M., and Gould-Somero, M. (1972): Segmental aneuploidy and the genetic gross structure of the *Drosophila* genome. *Genetics* 71:157–184.

Llinás, R., Steinberg, I. Z., and Walton, K. (1976): Presynaptic calcium currents and their relation to synaptic transmission: Voltage
clamp study in squid giant synapse and theoretical model for the calcium gate. *Proc. Nat. Acad. Sci.* 73:2918-2922.

Loh, H. H., Cho, T. M., Wu, Y. C., Harris, R. A., and Way, E. L.
(1975): Opiate binding to cerebroside sulfate: A model system for
opiate-receptor interaction. *Life Sci.* 16:1811-1818.

Low, P. A. (1977): The evolution of "onion bulbs" in hereditary
hypertrophic neuropathy of the Trembler mouse. *Neuropathol. Appl.
Neurobiol.* 3:81-92.

Low, P. A., and McLeod, J. G. (1975): Hereditary demyelinating
neuropathy in the trembler mouse. *J. Neurol. Sci.* 26:565-574.

Lunan, K. O., and Mitchell, H. K. (1969): The metabolism of tyro-
sine-O-phosphate in *Drosophila. Arch. Biochem. Biophys.* 132:
450-456.

Lund, R. D. (1978): *Development and Plasticity of the Brain.* New
York: Oxford University Press.

Macagno, E. R. (1978): Mechanism for the formation of synaptic projections in the arthropod visual system. *Nature* 275:318-320.

Macagno, E. R. (1980): Genetic approaches to invertebrate neurogenesis. In *Current Topics in Developmental Biology,* Vol. 15. *Neural
Development, Part 1. Emergence of Specificity in Neural Histogenesis.* Hunt, R. K., ed. New York: Academic Press, pp. 319-345.

MacBean, I. T., and Parsons, P. A. (1967): Directional selection for
duration of copulation in *Drosophila melanogaster. Genetics* 56:
233-239.

McCann, G. D. (1974): Nonlinear identification theory models for
successive stages of visual nervous systems of flies. *J. Neurophysiol.*
37:869-895.

McCann, G. D., and Dill, J. (1969): Fundamental properties of intensity, form and motion perception in the visual system nervous systems
of *Calliphora phaenicia* and *Musca domestica. J. Gen. Physiol.* 53:
385-413.

McClintock, B. (1967): Genetic systems regulating gene expression
during development. *Dev. Biol. Suppl.* 1:84-112.

McEwen, R. S. (1918): The reactions to light and to gravity in Drosophila and its mutants. *J. Exp. Zool.* 25:49-106.

McGuire, T. R., and Hirsch, J. (1977): Behavior-genetic analysis of

*Phormia regina:* Conditioning, reliable individual differences, and selection. *Proc. Nat. Acad. Sci.* 74:5193-5197.

McLaren, A. (1978): Sexual differentiation in mammalian chimeras and mosaics. In *Genetic Mosaics and Cell Differentiation,* Vol. 9. *Results and Problems in Cell Differentiation.* Gehring, W. J., ed. Berlin: Springer-Verlag, pp. 243-258.

MacLeod, A. R., Waterston, R. H., and Brenner, S. (1977): An internal deletion mutant of a myosin heavy chain in *Caenorhabditis elegans. Proc. Nat. Acad. Sci.* 74:5336-5340.

McMahon, D. (1974): Chemical messengers in development: A hypothesis. *Science* 185:1012-1021.

Macnab, R. M. (1978): Bacterial motility and chemotaxis: The molecular biology of a behavioral system. *CRC Crit. Rev. Biochem.* 5:291-341.

Macnab, R. M. (1980): Sensing the environment: bacterial chemotaxis. In *Biological Regulation and Development,* Vol. 2. Goldberger, R. F., ed. New York: Plenum Publishing Corp., pp. 377-412.

Macnab, R. M., and Koshland, D. E., Jr. (1972): The gradient-sensing mechanism in bacterial chemotaxis. *Proc. Nat. Acad. Sci.* 69: 2509-2512.

Maeda, K., and Imae, Y. (1979): Thermosensory transduction in *Escherichia coli:* Inhibition of the thermoresponse by L-serine. *Proc. Nat. Acad. Sci.* 76:91-95.

Malacinski, G. M., and Brothers, A. J. (1974): Mutant genes in the Mexican axolotl. *Science* 184:1142-1147.

Mallet, J. (1980): Biochemistry and immunology of neurological mutants in the mouse. In *Current Topics in Developmental Biology,* Vol. 15. *Neural Development. Part 1. Emergence of Specificity in Neural Histogenesis.* Hunt, R. K., ed. New York: Academic Press, pp. 41-65.

Manning, A. (1965): *Drosophila* and the evolution of behaviour. *Viewpoints Biol.* 4:123-167.

Manning, A. (1967a): The control of sexual receptivity in female *Drosophila. Animal Behav.* 15:239-250.

Manning, A. (1967b): Genes and the evolution of insect behavior. In *Behavior-Genetic Analysis.* Hirsch, J., ed. New York: McGraw-Hill, pp. 44-60.

Markow, T. A. (1981): Light-dependent pupation site preferences in

*Drosophila:* Behavior of adult visual mutants. *Behav. Neural Biol.* 31: 348–353.

Markow, T. A., and Manning, M. (1980): Mating success of photoreceptor mutants of *Drosophila melanogaster. Behav. Neural Biol.* 29:276–280.

Markow, T. A., and Merriam, J. (1977): Phototactic and geotactic behavior of countercurrent defective mutants of *Drosophila melanogaster. Behav. Genet.* 7:447–455.

Marler, P., and Peters, S. (1977): Selective vocal learning in a sparrow. *Science* 198:519–521.

Mato, J. M., and Marín-Cao, D. (1979): Protein and phospholipid methylation during chemotaxis in Dictyostelium discoideum and its relation to calcium movements. *Proc. Nat. Acad. Sci.* 76:6106–6109.

Meier, H., and MacPike, A. D. (1971): Three syndromes produced by two mutant genes in the mouse. Clinical, pathogloical, and ultrastructural bases of tottering, leaner, and heterozygous mice. *J. Hered.* 62: 297–302.

Meinertzhagen, I. A. (1973): Development of the compound eye and optic lobes of insects. In *Developmental Neurobiology of Arthropods.* Young, D., ed. Cambridge: Cambridge University Press, pp. 51–104.

Meinertzhagen, I. A. (1975): The development of neuronal connection patterns in the visual systems of insects. In *Cell Patterning.* (Ciba Foundation Symposium 29.) Amsterdam: Elsevier-Excerpta Medica-North-Holland, pp. 265–288.

Mendell, J. R., Higgins, R., Sahenk, Z., and Cosmos, E. (1979): Relevance of genetic animal models of muscular dystrophy to human muscular dystrophies. *Ann. N. Y. Acad. Sci.* 317:409–430.

Menne, D., and Spatz, H. C. (1977): Colour vision in *Drosophila melanogaster. J. Comp. Physiol.* 114:301–312.

Mergenhagen, D. (1980): Circadian rhythms in unicellular organisms. *Curr. Topics Microbiol. Immunol.* 90:123–147.

Merriam, J. R. (1978): Estimating primordial cell numbers in *Drosophila* imaginal discs and histoblasts. In *Genetic Mosaics and Cell Differentiation.* Gehring, W. J., ed. Berlin: Springer-Verlag, pp. 71–96.

Mesibov, R., and Adler, J. (1972): Chemotaxis toward amino acids in *Escherichia coli. J. Bacteriol.* 112:315–326.

Meyerowitz, E. M., and Kankel, D. R. (1978): A genetic analysis of visual system development in *Drosophila melanogaster. Dev. Biol.* 62: 112–142.

Millecchia, R., and Mauro, A. (1969): The ventral photoreceptor cells of *Limulus*. III. A voltage clamp study. *J. Gen. Physiol.* 54:331–351.

Miller, J. B., and Koshland, D. E., Jr. (1977): Sensory electrophysiology of bacteria: relationship of the membrane potential to motility and chemotaxis in *Bacillus subtilis*. *Proc. Nat. Acad. Sci.* 74: 4752–4756.

Mimura, K. (1971): Movement discrimination by the visual system of flies. *Z. Vergl. Physiol.* 73:105–138.

Mimura, K. (1976): Some spatial properties in the first optic ganglion of the fly. *J. Comp. Physiol.* 105:65–82.

Minke, B., Wu, C.-F., and Pak, W. L. (1975a): Induction of photoreceptor voltage noise in the dark in a *Drosophila* mutant. *Nature* 258: 84–87.

Minke, B., Wu, C.-F., and Pak, W. L. (1975b): Isolation of light-induced response of the central retinula cells from the electroretinogram of *Drosophila*. *J. Comp. Physiol.* 98:345–355.

Mintz, B. (1974): Genetic control of mammalian differentiation. *Ann. Rev. Genet.* 8:411–470.

Miquel, J. (1972): Aging of male *Drosophila melanogaster:* Histological, histochemical, and ultrastructural observations. *Adv. Gerontol. Res.* 3:39–71.

Morata, G., and Lawrence, P. A. (1977): Homeotic genes, compartments and cell determination in *Drosophila*. *Nature* 265:211–216.

Morata, G., and Ripoll, P. (1975): Minutes: Mutants of *Drosophila* autonomously affecting cell division rate. *Dev. Biol.* 42:211–221.

Morley, B. J., Kemp, G. E., and Salvaterra, P. (1979): $\beta$-Bungaratoxin binding sites in the CNS. *Life Sciences* 24:859–872.

Mullen, R. J. (1977a): Genetic dissection of the CNS with mutant-normal mouse and rat chimeras. In *Society for Neuroscience Symposia,* Vol. II. *Approaches to the Cell Biology of Neurons.* Cowan, W. M., and Ferrendelli, J. A., eds. Bethesda: Society for Neuroscience, pp. 47–65.

Mullen, R. J. (1977b): Site of *pcd* gene action and Purkinje cell mosaicism in cerebella of chimaeric mice. *Nature* 270:245–247.

Mullen, R. J. (1978): Mosaicism in the central nervous system of mouse chimeras. In *The Clonal Basis of Development.* (Thirty-sixth Symposium of the Society for Developmental Biology.) Subtelny, S., and Sussex, I. M., eds. New York: Academic Press, pp. 83–101.

Mullen, R. J., Eicher, E. M., and Sidman, R. L. (1976). Purkinje cell degeneration, a new neurological mutation in the mouse. *Proc. Nat. Acad. Sci.* 73:208–212.

Mullen, R. J., and Herrup, K. (1979): Chimeric analysis of mouse cerebellar mutants. In *Neurogenetics: Genetic Approaches to the Nervous System*. Breakfield, X. O., editor-in-chief. New York: Elsevier, pp. 173–196.

Mullen, R. J., and LaVail, M. M. (1975): Two new types of retinal degeneration in cerebellar mutant mice. *Nature* 258:528–530.

Mullen, R. J., and LaVail, M. M. (1976): Inherited retinal dystrophy: Primary defect in pigment epithelium determined with experimental rat chimeras. *Science* 192:779–801.

Muskavitch, M. A., Kart, E. N., Springer, M. S., Goy, M. F., and Adler, J. (1978): Attraction by repellents: An error in sensory information processing by bacterial mutants. *Science* 201:63–65.

Nakane, K. (1976): Postnatal development of brain in mice with congenital ataxia, rolling (*rol*) and tottering (*tg*). *Teratology* 14:248–249.

Narahashi, T. (1974): Chemicals as tools in the study of excitable membranes. *Physiol. Rev.* 54:813–889.

Nelson, D. L., and Kung, C. (1978): Behavior of Paramecium: Chemical, physiological and genetic studies. In *Taxis and Behaviour*. (Receptors and Recognition, Series B, Vol. 5.) Hazelbauer, G. L., ed. London: Chapman and Hall, pp. 75–100.

Nevers, P., and Saedler, H. (1977): Transposable genetic elements as agents of gene instability and chromosomal rearrangements. *Nature* 268:109–115.

Nissani, M. (1975): A new behavioral bioassay for analysis of sexual attraction and pheromones in insects. *J. Exp. Zool.* 192:271–275.

Nissani, M. (1977): Gynandromorph analysis of some aspects of sexual behaviour in *Drosophila melanogaster*. *Animal Behav.* 25: 555–556.

Noebels, J. L., and Sidman, R. L. (1979): Inherited epilepsy: Spike-wave and focal motor seizures in the mutant mouse tottering. *Science* 204:1334–1336.

Nyhan, W. L. (1973): The Lesch-Nyhan syndrome. *Ann. Rev. Med.* 24:41–60.

O'Brien, S. J., and MacIntyre, R. J. (1978): Genetics and biochemistry of enzymes and specific proteins of *Drosophila*. In *Genetics and Biology of Drosophila*, Vol. 2A. Ashburner, M., and Wright, R. R. F., eds. London: Academic Press, pp. 396–551.

O'Dea, R. F., Viveros, O. H., Axelrod, J., Aswanikumar, S., Schiffman, E., and Corcoran, B. A. (1978): Rapid stimulation of protein carboxymethylation in leukocytes by a chemotactic peptide. *Nature* 272:462–464.

Oertel, D., Schein, S. J., and Kung, C. (1977): Separation of membrane currents using a *Paramecium* mutant. *Nature* 268:120–124.

Ogura, A., and Takahashi, K. (1976): Artificial deciliation causes loss of calcium-dependent responses in *Paramecium*. *Nature* 264:170–172.

Ohno, S. (1979): *Major Sex Determining Genes*. Berlin: Springer-Verlag.

Olney, J. W., and deGubareff, T. (1978): Glutamate neurotoxicity and Huntington's chorea. *Nature* 271:557–559.

Ordal, G. W., and Adler, J. (1974): Properties of mutants in galactose taxis and transport. *J. Bacteriol.* 117:517–526.

Ostroy, S. E. (1978): The characteristics of *Drosophila* rhodopsin in wild-type and *norpA* vision transduction mutants. *J. Gen. Physiol.* 72:717–732.

Ostroy, S. E., and Pak, W. L. (1973): Protein differences associated with a phototransduction mutant of *Drosophila*. *Nature New Biol.* 243:120–121.

Ostroy, S. E., Wilson, M., and Pak, W. L. (1974): *Drosophila* rhodopsin: Photochemistry, extraction and differences in the *norpA*$^{P12}$ phototransduction mutant. *Biochem. Biophys. Res. Commun.* 59:960–966.

Ouweneel, W. J. (1976): Developmental genetics of homoeosis. *Adv. Genet.* 18:179–248.

Pak, W. L. (1975): Mutants affecting the vision of *Drosophila* melanogaster. In *Handbook of Genetics,* Vol. 3. King, R. C., ed. New York: Plenum Press, pp. 703–733.

Pak, W. L. (1979): Study of photoreceptor function using *Drosophila* mutants. In *Neurogenetics: Genetic Approaches to the Nervous System.* Breakefield, X. O., editor-in-chief. New York: Elsevier North-Holland, pp. 67–99.

Pak, W. L., Conrad, S. K., Kremer, N. E., Larrivee, D. C., Schinz, R. H., and Wong, F. (1980): Photoreceptor function. In *Development and Neurobiology of Drosophila.* Siddiqi, O., et al., eds. New York: Plenum Press, pp. 331–344.

Pak, W. L., Grossfield, J., and Arnold, K. S. (1970): Mutants of the visual pathway of *Drosophila melanogaster*. *Nature* 227:518–520.

Pak, W. L., and Lidington, K. J. (1974): Fast electrical potential from a long-lived, long-wavelength photoproduct of fly visual pigment. *J. Gen. Physiol.* 63:740–756.

Pak, W. L., Ostroy, S. E., Deland, M. C., and Wu, C.-F. (1976): Photoreceptor mutant of *Drosophila:* Is protein involved in intermediate steps of phototransduction? *Science* 194:956–959.

Pak, W. L., and Pinto, L. H. (1976): Genetic approach to the study of the nervous system. *Ann. Rev. Biophys. Bioengin.* 5:397–448.

Palka, J. (1977a): Abnormal neural development in invertebrates. In *Function and Formation of Neural Systems.* Stent, G. S., ed. Berlin: Dahlem Conference, pp. 139–159.

Palka, J. (1977b): Neurobiology of homeotic mutants in *Drosophila. Soc. Neurosci. Abstr.* 3:187.

Palka, J. (1979a): Mutants and mosaics: tools in insect developmental neurobiology. In *Society for Neuroscience,* Vol. IV. *Aspects of Developmental Neurobiology.* Ferrendelli, J. A., ed. Bethesda: Society for Neuroscience, pp. 209–227.

Palka, J. (1979b): Theories of pattern formation in insect neural development. *Adv. Insect. Physiol.* 14:251–349.

Palka, J., Lawrence, P. A., and Hart, H. G. (1979): Neural projection patterns from homeotic tissue of *Drosophila* studied in bithorax mutants and mosaics. *Dev. Biol.* 69:549–575.

Palka, J., and Schubiger, M. (1980): Formation of central patterns by receptor cell axons. In *Development and Neurobiology of Drosophila.* Siddiqi, O., et al., eds. New York: Plenum Press, pp. 223–246.

Parkinson, J. S. (1976): *cheA, cheB,* and *cheC* genes of *E. coli* and their role in chemotaxis. *J. Bacteriol.* 126:758–770.

Parkinson, J. S. (1977): Behavioral gentics in bacteria. *Ann. Rev. Genet.* 11:397–414.

Parkinson, J. S., and Parker, S. R. (1979): Interaction of the *cheC* and *cheZ* gene products is required for chemotactic behavior in *Escherichia coli. Proc. Nat. Acad. Sci.* 76:2390–2394.

Peterson, A. C. (1974): Chimaera mouse study shows absence of disease in genetically dystrophic muscle. *Nature* 248:561–564.

Peterson, A. C. (1979): Mosaic analysis of dystrophic ←→ normal chimeras: An approach to mapping the site of gene expression. *Ann. N. Y. Acad. Sci.* 317:630–648.

Peterson, A. C., Frair, P. M., Rayburn, H. R., and Cross, D. P. (1979): Development and disease in the neuromuscular system of muscular dystrophic ◄——► normal mouse chimeras. In *Society for Neuroscience Symposia,* Vol. IV. *Aspects of Developmental Neurobiology.* Ferrendelli, J. A., ed. Bethesda: Society for Neuroscience, pp. 258-273.

Pfeffer, W. (1884): Locomotorische Richtungsbewegungen durch chemische Reize. *Unters. Bot. Inst. Tübingen* 1:363-482.

Pierantoni, R. (1976): A look into the cock-pit of the fly. Th architecture of the lobular plate. *Cell Tiss. Res.* 171:101-122.

Pike, M. C., Kredich, N. M., and Snyderman, R. (1978): Requirement of S-adenosyl-L-methionine-mediated methylation for human monocyte chemotaxis. *Proc. Nat. Acad. Sci.* 75:3928-3932.

Pitman, R. M. (1971): Transmitter substances in insects: A review. *Comp. Gen. Pharmacol.* 2:347-371.

Pittendrigh, C. S. (1967): Circadian systems. I. The Driving oscillation and its assay in *Drosophila pseudoobscura. Proc. Nat. Acad. Sci.* 58:1762-1767.

Pittendrigh, C. S. (1974): Circadian oscillations in cells and the circadian organization of multicellular systems. In *The Neurosciences: Third Study Program.* Schmitt, F. O., and Worden, F. G., editors-in-chief. Cambridge, MA: MIT, pp. 437-458.

Pittendrigh, C. S., and Minis, D. H. (1971): The photoperiodic time measurement in *Pectinophora gossypiella* and its relation to the circadian system in that species. In *Biochronometry.* Menaker, M., ed. Washington, D. C.: National Academy of Sciences, pp. 212-250.

Plum, F., Gjedde, A., and Samson, F. E. (1976): Neuroanatomical functional mapping by the radioactive 2-deoxy-D-glucose method. *Neurosci. Res. Prog. Bull.* 14:457-518.

Pollard, J. D. (1978): Nerve allografts in trembler mice. In *Fourth International Congress on Neuromuscular Diseases.* Abstr. 8.

Poodry, C. A., and Edgar, L. (1979): Reversible alterations in the neuromuscular junctions of *Drosophila melanogaster* bearing a temperature-sensitive mutation, *shibire. J. Cell Biol.* 81:520-527.

Poodry, C. A., Hall, L., and Suzuki, D. T. (1973): Developmental properties of *shibire^(ts1).* A pleiotropic mutation affecting larval and adult locomotion and development. *Dev. Biol.* 32:373-386.

Potter, S. S., and Thomas, C. A., Jr. (1978): The two-dimensional fractionation of *Drosophila* DNA. *Cold Spring Harbor Symp. Quant. Biol. Part 2. Chromatin.* 42:1023-1031.

Poulson, D. F. (1950): Histogenesis, organogenesis, and differentiation in the embryo of *Drosophila melanogaster* Meigen. In *Biology of Drosophila*. Demerec, M., ed. New York: John Wiley, pp. 168–274.

Power, M. E. (1943): The brain of *Drosophila melanogaster*. *J. Morphol.* 72:517–559.

Preer, J. R., Jr. (1968): Genetics of the protozoa. In *Research in Protozoology*. Chen, T. T., ed. New York: Pergamon Press, pp. 139–288.

Preston, M. S., Guthrie, J. T., and Childs, B. (1974): Visual evoked responses in normal and disabled readers. *Psychophysiology* 11: 452–457.

Preston, M. S., Guthrie, J. T., Kirsch, I., Gertman, D., and Childs, B. (1977): VERs in normal and disabled adult readers. *Psychophysiology* 14:8–13.

Pye, K., and Chance, B. (1966): Sustained sinusoidal oscillations of reduced pyridine nucleotide in a cell-free extract of *Saccharomyces carlsbergensis*. *Proc. Nat. Acad. Sci.* 55:888–894.

Quinn, W. G., and Dudai, Y. (1976): Memory phases in *Drosophila*. *Nature* 262:576–577.

Quinn, W. G., Harris, W. A., and Benzer, S. (1974): Conditioned behavior in *Drosophila melanogaster*. *Proc. Nat. Acad. Sci.* 71: 708–712.

Quinn, W. G., Sziber, P. P., and Booker, R. (1979): The *Drosophila* memory mutant *amnesiac*. *Nature* 277:212–214.

Quinn, T. C., and Craig, G. B. (1971): Phenogenetics of the homeotic mutant *proboscipedia* in *Aedes albopictus*. *J. Hered.* 62:3–12.

Raisman, G., and Field, P. M. (1973): Sexual dimorphism in the neuropil of the preoptic area of the rat and its dependence on neonatal androgen. *Brain Res.* 54:1–29.

Rakic, P. (1975): Synaptic specificity in the cerebellar cortex: Study of anomalous circuits induced by single gene mutations in mice. *Cold Spring Harbor Symp. Quant. Biol.* 40:333–346.

Rakic, P., and Sidman, R. L. (1973): Weaver mutant mouse cerebellum: Defective neuronal migration secondary abnormality of Bergmann glia. *Proc. Nat. Acad. Sci.* 70:240–244.

Rakic, P., Stensaas, L. J., Sayre, E. P., and Sidman, R. L. (1974): Computer-aided three-dimensional reconstruction and quantitative analysis of cells from serial electron microscopic montages of foetal monkey brain. *Nature* 250:31–34.

Rathbone, M. P., Stewart, P. A., and Vetrano, F. (1975): Dystrophic spinal cord transplants induce abnormal thymidine kinase activity in normal muscles. *Science* 189:1106–1107.

Razzak, A., Fujiwara, M., and Veki, J. (1975): Automutilation induced by clonidine in mice. *Eur. J. Pharmacol.* 30:356–359.

Ready, D. F., Hanson, T. E., and Benzer, S. (1976): Development of the *Drosophila* retina, a neurocrystalline lattice. *Dev. Biol.* 53: 217–240.

Reichardt, W. E. (1970): The insect eye as a model for analysis of uptake, transduction, and processing of optical data in the nervous system. In *Neurosciences: Second Study Program.* Schmitt, F. O., editor-in-chief. New York: Rockefeller University Press, pp. 494–511.

Reichardt, W. E., and Poggio, T. (1976): Visual control of orientation behavior in the fly. Part I. A quantitative analysis. *Quart. Rev. Biophys.* 9:311–375.

Riddle, D. L. (1977): A genetic pathway for dauer larva formation in *Caenorhabditis elegans. Stadler Genet. Symp.* 9:101–120.

Riddle, D. L. (1978): The genetics of development and behavior in *Caenorhabditis elegans. J. Nematol.* 10:1–16.

Riddle, D. L., Swanson, M. M., and Albert, P. S. (1981): Interacting genes in nematode dauer larva formation. *Nature* 290:668–671.

Riddle, D. L. (1980): Developmental genetics of *Caenorhabditis elegans.* In *Nematodes as Biological Models,* Vol. 1. Zuckerman, B. M., ed. New York: Academic Press, pp. 263–283.

Riddle, D. L., and Brenner, S. (1978): Indirect suppression in *Caenorhabditis elegans. Genetics* 89:299–314.

Rizki, T. M., and Rizki, R. M. (1978): Larval adipose tissue of homoeotic mutants of *Drosophila. Dev. Biol.* 65:476–482.

Rodrigues, V. (1980): Olfactory behavior of *Drosophila melanogaster.* In *Development and Neurobiology of Drosophila.* Siddiqi, O., et al., eds. New York: Plenum Press, pp. 361–369.

Rodrigues, V., and Siddiqi, O. (1978): Genetic analysis of chemosensory pathway. *Proc. Indian Acad. Sci.* 87B:147–160.

Roeder, K. D. (1967): *Nerve Cells and Insect Behavior.* Cambridge, MA: Harvard University Press.

Roffler-Tarlov, S., Beart, P. M., O'Gorman, S., and Sidman, R. L. (1979): Neurochemical and morphological consequences of axon terminal degeneration in cerebellar deep nuclei of mice with inherited Purkinje cell degeneration. *Brain Res.* 168:75–95.

Roffler-Tarlov, S., and Sidman, R. L. (1978): Concentrations of glutamic acid in cerebellar cortex and deep nuclei of normal mice and weaver, staggerer and nervous mutants. *Brain Res.* 142:269–283.

Romine, J. S., Bray, G. M., and Aguayo, A. J. (1976): Schwann cell multiplication after crush injury of unmyelinated fibers. A radioautographic and electron microscopical study. *Arch. Neurol.* 33:49–54.

Rosenbloom, F. M., Kelley, W. N., Miller, J. M., Henderson, J. F., and Seegmiller, J. E. (1967): Inherited disorder of purine metabolism: Correlation between central nervous system dysfunction and biochemical defects. *J. Am. Med.* 202:175–177.

Ross, M. H. (1964): Prenatal wings in Blattella germanica and their possible evolutionary significance. *Am. Midland Naturalist* 71: 161–180.

Rotmann, E. (1939): Der Anteil von Inductor und reagierendem Gewebe an der Entwicklung der Amphibienlinse. *Wilhelm Roux' Arch. Entwicklungsmech. Org.* 139:1–49.

Rubik, B. A., and Koshland, D. E. (1978): Potentiation, desensitization, and inversion of response in bacterial sensing of chemical stimuli. *Proc. Nat. Acad. Sci.* 75:2820–2824.

Rubin, G. M., Brorein, W. J., Dunsmuir, P., Levis, R., Potter, S. S., Strobel, E., and Young, E. (1980): Transposable elements in the *Drosophila* genome. In *Mobilization and Reassembly of Genetic Information.* Scott, W. A., Werner, R., Joseph, D. R., and Schultz, J., eds. New York: Academic Press, pp. 235–242.

Rudloff, E., Jimenez, F., and Bartels, J. (1980): Purification and properties of the nicotinic acetylcholine receptor of *Drosophila melanogaster.* In *Receptors for Neurotransmitters, Hormones, and Pheromones in Insects.* Satelle, D. B., Hall, L. M., and Hildebrand, J. G., eds. Amsterdam: Elsevier/North-Holland Biomedical Press, pp. 85–92.

Russell, R. L., Johnson, C. D., Rand, J. B., Scherer, S., and Zwass, M. S. (1977): Mutants of acetylcholine metabolism in the nematode *Caenorhabditis elegans.* In *Molecular Approaches to Eucaryotic Genetic Systems.* Wilcox, G., Abelson, J., and Fox, C. F., eds. New York: Academic Press, pp. 359–371.

Sachar, E. J., and Baron, M. (1979): The biology of affective disorders. *Ann. Rev. Neurosci.* 2:505–518.

Saimi, Y., and Kung, C. (1980): A Ca-induced Na-current in *Paramecium. J. Exp. Biol.* 88:305–326.

Salkoff, L., and Kelly, L. (1978): Temperature-induced seizure and frequency-dependent neuromuscular block in a *ts* mutant of *Drosophila*. *Nature* 273:156–158.

Salkoff, L., and Kelly, L. (1980): Amino-pyridines mimic mutant *Drosophila* developmental defects. *Comp. Biochem. Physiol.* 65: 59–63.

Samorajski, T., Friede, R. L., and Reimer, P. R. (1970): Hypomyelination in the quaking mouse. A model for the analysis of disturbed myelin formation. *J. Neuropathol. Exp. Neurol.* 29:507–523.

Sanderson, K. J., Guillery, R. W., and Shackelford, R. M. (1974): Congenitally abnormal visual pathways in mink (Mustela vision) with reduced retinal pigment. *J. Comp. Neurol.* 154:225–248.

Sanyal, S., and Zeilmaker, G. H. (1977): Cell lineage in retinal development of mice studied in experimental chimaeras. *Nature* 265: 731–733.

Sargent, M. L., Briggs, W. R., and Woodward, D. O. (1966): The circadian nature of a rhythm expressed by an invertaseless strain of Neurospora crassa. *Plant Physiol.* 41:1343–1349.

Satelle, D. M., Hall, L. M., and Hildebrand, J. G., eds. (1980): *Receptors for Neurotransmitters, Hormones and Pheromones in Insects*. Amsterdam: Elsevier/North-Holland Biomedical Press.

Satow, Y. (1981): Membrane excitability in protozoan—*Paramecium tetraurelia*. In *Invertebrate Membrane Physiology*. Podesta, R. B., ed. New York: Marcel Dekker (in press).

Satow, Y., Chang, S. Y., and Kung, C. (1974): Membrane excitability: Made temperature-dependent by mutations. *Proc. Nat. Acad. Sci.* 71:2703–2706.

Satow, Y., and Kung, C. (1976a): A mutant of *Paramecium* with increased relative resting potassium permeability. *J. Neurobiol.* 7: 325–338.

Satow, Y., and Kung, C. (1976b): Mutants with reduced Ca activation in *Paramecium aurelia*. *J. Mem. Biol.* 28:277–294.

Satow, Y., and Kung, C. (1976c): 'TEA$^+$-insensitive' mutant with increased potassium conductance in *Paramecium aurelia*. *J. Exp. Biol.* 65:51–63.

Satow, Y., and Kung, C. (1980): Membrane currents of pawn mutants of the *pwA* group in *Paramecium tetraurelia*. *J. Exp. Biol.* 84:57–71.

Saunders, D. S. (1976): *Insect Clocks*. New York: Pergamon Press.

Schein, S. J. (1976a): Calcium channel stability measured by gradual loss of excitability in pawn mutants of *Paramecium aurelia. J. Exp. Biol.* 65:725–736.

Schein, S. J. (1976b): Nonbehavioral selection for pawns, mutants of *Paramecium aurelia* with decreased excitability. *Genetics* 84:453–468.

Schein, S. J., Bennett, M. V. L., and Katz, G. M. (1976): Altered calcium conductance in pawns, behavioral mutants of *Paramecium aurelia. J. Exp. Biol.* 65:699–724.

Schilcher, F. V. (1976a): The behavior of cacophony, a courtship song mutant in *Drosophila melanogaster. Behav. Biol.* 17:187–196.

Schilcher, F. V. (1976b): The function of pulse song and sine song in the courtship of *Drosophila melanogaster. Animal Behav.* 24: 622–625.

Schilcher, F. V. (1977): A mutation which changes courtship song in *Drosophila melanogaster. Behav. Genet.* 7:251–259.

Schilcher, F. V., and Hall, J. C. (1979): Neural topography of courtship song in sex mosaics of *Drosophila melanogaster. J. Comp. Physiol.* 129:85–95.

Schilcher, F. V., and Manning, A. (1975): Courtship and mating speed in hybrids between *Drosophila melanogaster* and *Drosophila simulans. Behav. Genet.* 5:395–404.

Schinz, R. H. (1978): Cell junctions between the photoreceptor cells of *Drosophila. Soc. Neurosci. Abstr.* 4:248.

Schmidt-Nielsen, B. K., Gepner, J. I., Teng, N. N. H., and Hall, L. M. (1977): Characterization of an $\alpha$-bungarotoxin binding component from *Drosophila melanogaster. J. Neurochem.* 29: 1013–1029.

Schwartz, W. J., Davidsen, L. C., and Smith, C. B. (1980): *In vivo* metabolic activity of a putative circadian oscillator, the rat suprachiasmatic nucleus. *J. Comp. Neurol.* 189:157–167.

Sebald, W., and Wachter, E. (1978): Amino acid sequence of the putative protonophore of the energy-transducing ATPase complex. In *Energy Conservation in Biological Membranes.* Schäfer, G., and Klingenberg, M., eds. Berlin: Springer-Verlag, pp. 228–336.

Seegmiller, J. E. (1976): Inherited deficiency of hypoxanthine-guanine phosphoribosyltransferase in X-linked uric acidura (the Lesch-Nyhan syndrome and its variants). *Adv. Human Genet.* 6:75–163.

Seegmiller, J. E., Rosenbloom, F. M., and Kelley, W. N. (1967): Enzyme defect associated with a sex-linked human neurological disorder and excessive purine synthesis. *Science* 155:1682–1684.

Seil, F. J., Blank, N. K., and Leiman, A. L. (1979): Toxic effects of kainic acid on mouse cerebellum in tissue culture. *Brain Res.* 161: 253–265.

Shatz, C. J. (1977a): Abnormal interhemispheric connections on the visual system of Boston Siamese cats: A physiological study. *J. Comp. Neurol.* 171:229–246.

Shatz, C. J. (1977b): Anatomy of interhemispheric connections in the visual system of Boston Siamese and ordinary cats. *J. Comp. Neurol.* 173:497–518.

Shatz, C. J. (1977c): A comparison of visual pathway in Boston and Midwestern Siamese cats. *J. Comp. Neurol.* 171:205–228.

Shatz, C. J. (1977d): Normal and abnormal organization of the mammalian primary visual cortex. *Biosci. Commun.* 3:413–434.

Shatz, C. J., and LeVay, S. (1979): Siamese cat: altered connections of visual cortex. *Science* 204:328–330.

Shellenbarger, D. L., and Mohler, J. D. (1978): Temperature-sensitive periods and autonomy of pleiotropic effects of *l(1) N$^{ts1}$,* a conditional Notch lethal in *Drosophila. Dev. Biol.* 62:432–446.

Shorey, H. H., and Bartell, R. J. (1970): Role of a volatile female sex pheromone in stimulating male courtship behavior in *Drosophila melanogaster. Animal Behav.* 18:159–164.

Sibitani, A. (1980): Wing homeosis in Lepidoptera: A survey. *Dev. Biol.* 79:1–18.

Siddiqi, O. (1975): Genetic blocks in the elements of neural network in *Drosophila.* In *Regulation of Growth and Differentiated Function in Eukaryote Cells.* Talwar, G. P., ed. New York: Raven Press, pp. 541–549.

Siddiqi, O., and Benzer, S. (1976): Neurophysiological defects in temperature-sensitive paralytic mutants of *Drosophila melanogaster. Proc. Nat. Acad. Sci.* 73:3253–3257.

Siddiqi, O., and Rodrigues, V. (1980): Genetic analysis of a complex chemoreceptor. In *Development and Neurobiology of Drosophila.* Siddiqi, O., et al., eds. New York: Plenum Press, pp. 347–359.

Siddiqui, S. S., and Babu, P. (1980): Genetic mosaics of *Caenorhabditis elegans:* A tissue-specific fluorescent mutant. *Science* 210: 330–332.

Sidman, R. L. (1973): Cell-cell recognition in the developing central nervous system. In *Neurosciences: Third Study Program.*

Schmitt, F. O., and Worden, F. G., Editors-in-chief. Cambridge, MA: MIT Press, pp. 743-758.

Sidman, R. L. (1974): Contact interaction among developing mammalian brain cells. In *The Cell Surface in Development*. Moscona, A., ed. New York: John Wiley, pp. 221-253.

Sidman, R. L. (1975): Cell interaction in mammalian brain development. In *The Nervous System,* Vol. 1. *Basic Neurosciences.* Tower, D. B., editor-in-chief. New York: Raven Press, pp. 601-610.

Sidman, R. L. (1976): Cell surface properties and the expression of inherited brain diseases in mice. In *Membranes and Disease*. Bolis, L., et al., eds. New York: Raven Press, pp. 379-386.

Sidman, R. L. (1977): Some rules governing specificity of synaptic connections in the developing mammalian brain. In *Research to Practice in Mental Retardation. Biomedical Aspects.* (International Association for the Scientific Study of Mental Deficiency, IASSMD), 3 Vols. Mittler, P., ed. Baltimore: University Park Press, pp. 375-389.

Sidman, R. L. (1978): Epigenetic sequences in CNS Development. *J. Neuropathol. Exp. Neurol.* 37:571.

Sidman, R. L., Angevine, J. B., Jr., and Pierce, E. T. (1971): *Atlas of the Mouse Brain and Spinal Cord*. Cambridge, MA: Harvard University Press.

Sidman, R. L., Cowen, J. S., and Eicher, E. M. (1979): Inherited muscle and nerve diseases in mice: a tabulation with commentary. *Ann. N. Y. Acad. Sci.* 317:497-505.

Sidman, R. L., Dickie, M. M., and Appel, S. H. (1964): Mutant mice (quaking and jumpy) with deficient myelination in the central nervous system. *Science* 144:309-311.

Sidman, R. L., Green, M. C., and Appel, S. H. (1965): *Catalog of the Neurological Mutants of the Mouse.* Cambridge, MA: Harvard University Press.

Sidman, R. L., and Rakic, P. (1973): Neuronal migration, with special reference to developing human brain: A review. *Brain Res.* 62: 1-35.

Sidman, R. L., and Rakic, P. (1975): Development of the human central nervous system. In *Cytology and Cellular Pathology of the Nervous System*. Haymaker, W., and Adams, R. D., eds. Springfield, IL: C. C. Thomas.

Sidman, R. L., and Wessells, N. K. (1975): Control of direction of growth during the elongation of neurites. *Exp. Neurol.* 48:237-251.

Siegel, R. W., and Hall, J. C. (1979): Conditioned responses in courtship behavior of normal and mutant *Drosophila*. *Proc. Nat. Acad. Sci.* 76:3430-3434.

Silman, I., and Dudai, Y. (1975): Acetylcholinesterases. In *Research Methods in Neurochemistry,* Vol. 3. Marks, N., and Rodnight, R., eds. New York: Plenum Press, pp. 209-252.

Silver, J., and Hughes, A. F. W. (1974): The relationship between morphogentic cell death and the development of congenital anophthalmia. *J. Comp. Neurol.* 157:281-302.

Silver, J., and Robb, R. M. (1979): Studies on the development of the eye cup and optic nerve in normal mice and in mutants with congenital optic nerve aplasia. *Dev. Biol.* 68:175-190.

Silver, J., and Sidman, R. L. (1980): A mechanism for the guidance and topographic patterning of retinal ganglion cell axons. *J. Comp. Neurol.* 189:101-111.

Silverman, M., and Simon, M. (1974): Flagellar rotation and the mechanism of bacterial motility. *Nature* 249:73-74.

Silverman, M., and Simon, M. (1977a): Chemotaxis in *Escherichia coli:* Methylation of *che* gene products. *Proc. Nat. Acad. Sci.* 74: 3317-3321.

Silverman, M., and Simon, M. (1977b): Identification of polypeptides necessary for chemotaxis in *Escherichia coli. J. Bacteriol.* 130: 1317-1325.

Simchen, G. (1978): Cell cycle mutants. *Ann. Rev. Genet.* 12: 161-191.

Skolnick, P., Syapin, P. J., Paugh, B. A., and Paul, S. M. (1979): Reduction in benzodiazepine receptors associated with Purkinje cell degeneration in 'nervous' mutant mice. *Nature* 277:397-399.

Snyder, M. A., Stock, J. B., and Koshland, D. E., Jr. (1981): Role of membrane potential and calcium in chemotactic sensing by bacteria. *J. Mol. Biol.* 149:241-257.

Snyder, S. H. (1978): Dopamine and schizophrenia. In *The Nature of Schizophrenia. New Approaches to Research and Treatment.* Wynne, L. C., Cromwell, R. L., and Matthysse, S., eds. New York: John Wiley, pp. 87-94.

Sokoloff, L., Reivich, M., Kennedy, C., DesRosiers, M. H., Patlak, K. D., and Pettigrew, O. (1977): The [$^{14}$C] deoxyglucose method for the measurement of local cerebral glucose utilization: Theory, procedure, and normal values in the conscious and anesthetized albino rat. *J. Neurochem.* 28:897-916.

Søndergaard, L. (1976): Temperature-induced changes in succinate-cytochrome *c* reductase in behavioural mutants of *Drosophila*. *Hereditas* 82:51–56.

Søndergaard, L. (1980): Dominant cold puralytic mutations on the autosomes of *Drosophila* melanogaster. *Hereditas* 92:335–340.

Sotelo, C., and Changeux, J.-P. (1974): Bergmann fibers and granular cell migration in the cerebellum of homozygous *weaver* mutant mouse. *Brain Res.* 77:484–491.

Spatz, H. C., Emanns, A., and Reichert, H. (1974): Associative learning of *Drosophila melanogaster. Nature* 248:359–361.

Spemann, H. (1918): Über die Determination der ersten Organanlagen des Amphibienembryo I-VI. *Wilhelm Roux' Arch. Entwicklungsmech. Org.* 43:448–555.

Spemann, H., and Mangold, H. (1924): Über Induktion von Embryonalanlagen durch Implantation artfremder Organisatoren. *Arch. Mikrosk. Anat.* 100:599–638.

Springer, M. S., Goy, M. F., and Adler, J. (1977): Sensory transduction in *Escherichia coli:* Two complementary pathways of information processing that involve methylated proteins. *Proc. Nat. Acad. Sci.* 74: 3312–3316.

Springer, M. S., Goy, M. F., and Adler, J. (1979): Protein methylation in behavioural control mechanisms and in signal transduction. *Nature* 280:279–284.

Springer, M. S., Kort, E. N., Larsen, S. H., et al. (1975): Role of methionine bacterial chemotaxis: Requirement for tumbling and involvement in information processing. *Proc. Nat. Acad. Sci.* 72: 4640–4644.

Stadler, D. R. (1958): Genetic control a cyclic growth pattern in *Neurospora. Nature* 184:170–171.

Steel, K. P., and Bock, G. R. (1980): The nature of inherited deafness in *deafness* mice. *Nature* 288:159–161.

Stern, C. (1969): Somatic recombination within the white locus of *Drosophila melanogaster. Genetics* 62:573–581.

Stewart, M., Murphy, C., and Fristrom, J. W. (1972): The recovery and preliminary characterization of X chromosome mutants affecting imaginal discs of *Drosophila melanogaster. Dev. Biol.* 27:71–83.

Stirling, C. A. (1975): Abnormalities in Schwann cell sheaths in spinal nerve roots of dystrophic mice. *J. Anat.* 119:169–180.

Stocker, R. F. (1977): Gustatory stimulation of a homeotic mutant appendage, *Antennapedia,* in *Drosophila melanogaster. J. Comp. Physiol.* 115:351-361.

Stocker, R. F. (1979): Fine structural comparison of the antennal nerve in the homeotic mutant *Antennapedia* with the wild-type antennal and second leg nerves of *Drosophila melanogaster. J. Morphol.* 160:209-222.

Stocker, R. F., Edwards, J. S., Palka, J., and Schubiger, G. (1976): Projection of sensory neurons from a homeotic mutant appendage, *Antennapedia,* in *Drosophila melanogaster. Dev. Biol.* 52:210-220.

Stocker, R. F., and Lawrence, P. A. (1981): Sensory projections from normal and homoeotically transformed antennae in *Drosophila. Dev. Biol.* 82:224-237.

Stone, J., Campion, J. E., and Leicester, J. (1978): The nasotemperal division of retina in the Siamese cat. *J. Comp. Neurol.* 180:783-798.

Stone, L. S. (1980): Heteroplastic transplantation of eyes between the larvae of two species of *Amblystoma. J. Exp. Zool.* 55:193-261.

Strange, P. G., and Koshland, D. E. (1976): Receptor interactions in a signalling system: Competition between ribose receptor and galactose receptor in the chemotaxis response. *Proc. Nat. Acad. Sci.* 73: 762-766.

Strausfeld, N. J. (1976): *Atlas of an Insect Brain.* New York: Springer-Verlag.

Strausfeld, N. J. (1980): Male and female visual neurones in dipterous insects. *Nature* 283:381-383.

Strausfeld, N. J., and Braitenberg, V. (1970): The compound eye of the fly (*Musca domestica*): Connections between the cartridges of the lamina ganglionaris. *Z. Vergl. Physiol.* 70:95-104.

Strausfeld, N. J., and Obermayer, M. (1976): Resolution of intra-neuronal and transynaptic migration of cobalt in the insect visual and central nervous system. *J. Comp. Physiol.* 110:1-12.

Strausfeld, N. J., and Singh, R. N. (1980): Peripheral and central nervous system projections in normal and mutant (bithorax) *Drosophila melanogaster.* In *Development and Neurobiology of Drosophila.* Siddiqi, O., et al., eds. New York: Plenum Press, pp. 267-290.

Stretton, A. O. W., Fishpool, R. M., Southgate, E., Donmoyer, J. E., Walrond, J. P., Moses, J. E. R., and Kass, I. S. (1978): Structure and physiological activity of the motoneurons of the nematode *Ascaris. Proc. Nat. Acad. Sci.* 75:3493-3497.

Strumwasser, F. (1965): The demonstration and manipulation of a circadian rhythm in a single neuron. In *Circadiam Clocks*. Aschoff, J., ed. Amsterdam: North Holland, pp. 442–462.

Sturtevant, A. H. (1915): Experiments on sex recognition and the problem of sexual selection in *Drosophila. J. Animal Behav.* 5: 351–366.

Sulston, J. E. (1976): Post-embryonic development in the ventral cord of *Caenorhabditis elegans. Phil. Trans. Roy. Soc. B* 275:287–297.

Sulston, J. E., Dew, M., and Brenner, S. (1975): Dopaminergic neurons in the nematode *Caenorhabditis elegans. J. Comp. Neurol.* 163: 215–226.

Sulston, J. E., and Horvitz, H. R. (1977): Post-embryonic cell lineages of the nematode, *Caenorhabditis elegans. Dev. Biol.* 56:110–156.

Sulston, J. E., and Horvitz, H. R. (1981): Abnormal cell lineages in mutants of the nematode *Caenorhabditis elegans. Dev. Biol.* 82: 41–55.

Suzuki, D. T., Grigliatti, T., and Williamson, R. (1971): Temperature-sensitive mutations in *Drosophila melanogaster*. VII. A mutation (para[ts]) causing reversible adult paralysis. *Proc. Nat. Acad. Sci.* 68: 890–893.

Swanson, M. M., and Poodry, C. A. (1980): Pole cell formation in *Drosophila melanogaster. Dev. Biol.* 75:419–430.

Szmelcman, S., and Adler, J. (1976): Change in membrane potential during bacterial chemotaxis. *Proc. Nat. Acad. Sci.* 73:4387–4391.

Takahashi, M. (1979): Behavioral mutants of *Paramecium caudatum. Genetics* 91:393–408.

Takahashi, M., and Naitoh, Y. (1978): Behavioral mutants of *Paramecium caudatum* with defective membrane electrogenesis. *Nature* 271:656–658.

Tanimura, T., Isono, K., and Kikuchi, T. (1978): Partial "sweet taste blindness" and configurational requirement of stimulants in a *Drosophila* mutant. *Jpn. J. Genet.* 53:71–73.

Tauc, L., and Bruner, J. (1963): "Desensitization" of cholinergic receptors by acetylcholine in molluscan central neurones. *Nature* 198: 33–34.

Tazima, Y. (1964): *The Genetics of the Silkworm Moth*. New York: Academic Press.

Teh, J. Z., Oxford, G. S., Wu, C. H., and Narahoshi, T. (1976): Dynamics of aminopyridine block of potassium channels in squid axon membrane. *J. Gen. Physiol.* 68:519–535.

Theiler, K., Varnum, D. S., Nadeau, J. H., Stevens, L. C., and Cagianut, B. (1976): A new allele of ocular retardation: Early development and morphogenetic cell death. *Anat. Embryol.* 150:85–97.

Thesleff, S. (1980): Aminopyridines and synaptic transmission. *Neuroscience* 5:1413–1419.

Thorpe, W. H. (1958): Further studies on the process of song learning in the chaffinch (*Fringilla coelebs gengleri*). *Nature* 182:554–557.

Toews, M. L., Goy, M. F., Springer, M. S., and Adler, J. (1979): Attractants and repellents control demethylation of methylated chemotaxis proteins in *Escherichia coli. Proc. Nat. Acad. Sci.* 76: 5544–5548.

Tomkins, G. M. (1975): The metabolic code. *Science* 189:760–768.

Tompkins, L. (1979): Developmental analysis of two mutations affecting chemotactic behavior in *Drosophila melanogaster. Dev. Biol.* 73:174–177.

Tompkins, L., Cardosa, M. J., White, F. V., and Sanders, T. G. (1979): Isolation and analysis of chemosensory behavior mutants in *Drosophila melanogaster. Proc. Nat. Acad. Sci.* 76:884–886.

Tompkins, L., Gross, A., Hall, J. C., Gailey, D. A., and Siegel, R. W. (1981): Genetic analysis of the role of female movement in sexual behavior of *Drosophila melanogaster. Behav. Genet.* (in press).

Tompkins, L., and Hall, J. C. (1981): The different effects on courtship of volatile compounds from mated and virgin *Drosophila* females. *J. Insect Physiol.* 271:17–21.

Tompkins, L., Hall, J. C., and Hall, L. M. (1980): Courtship-stimulating volatile compounds from normal and mutant *Drosophila. J. Insect Physiol.* 29:689–697.

Trenkner, E., Hatten, M. E., and Sidman, R. L. (1976): Histogenesis of normal and mutant mouse cerebellar cells in a microculture system. *Soc. Neurosci. Abstr.* 2:1028.

Trenkner, E., Hatten, M. E., and Sidman, R. L. (1978): Effect of ether-soluble serum components *in vitro* on the behavior of immature cerebellar cells in weaver mutant mice. *Neuroscience* 3:1093–1100.

Trenkner, E., and Sidman, R. L. (1976): Histogenesis of mouse cerebellum in microwell cultures: Cell reaggregation and migration, fiber and synapse formation. *J. Cell. Biol.* 75:915–940.

Trout, W. E., and Kaplan, W. D. (1970): A relationship between longevity, metabolic rate, and activity in shaker mutant of *Drosophila melanogaster. Exp. Gerontol.* 5:83–92.

Trout, W. E., and Kaplan, W. D. (1973): Genetic manipulation of motor output in shaker mutants of *Drosophila*. *J. Neurobiol*. 4: 495–512.

Tsuji, S., and Meier, H. (1971): Evidence for allelism of leaner and tottering in the mouse. *Genet. Res*. 17:83–88.

Tunnicliff, G., Rick, J. T., and Connolly, K. (1969): Locomotor activity in Drosophila. V. A comparative biochemical study of selectively bred populations. *Comp. Biochem. Physiol*. 29:1239–1245.

Twitty, V. C. (1930): Regulation in the growth of transplanted eyes. *J. Exp. Zool*. 55:43–52.

Twitty, V. C. (1932): Influence of the eye on the growth of its associated structures, studied by means of heteroplastic transplantation. *J. Exp. Zool*. 61:333–374.

Twitty, V. C. (1937): Experiments on the phenomenon of paralysis produced by a toxin occurring in Triturus embryos. *J. Exp. Zool*. 76: 67–104.

Twitty, V. C. (1940): Size-controlling factors. *Growth (Suppl.)* 4: 109–120.

Twitty, V. C. (1955): Eye. In *Analysis of Development*. Willier, B. H., Weiss, P. A., and Hamburger, V., eds. Philadelphia: W. B. Saunders, pp. 402–414.

Twitty, V. C. (1966): *Of Scientists and Salamanders*. San Francisco: W. H. Freeman.

Twitty, V. C., and Elliott, H. A. (1934): The relative growth of the amphibian eye, studied by means of transplantation. *J. Exp. Zool*. 68: 247–291.

Twitty, V. C., and Schwind, J. L. (1931): The growth of eyes and limbs transplanted heteroplastically between two species of *Amblystoma*. *J. Exp. Zool*. 59:61–86.

Ulbricht, W. (1977): Ionic channels and gating currents in excitable membranes. *Ann. Rev. Biophys. Bioeng*. 6:7–31.

Ulshafer, R. J., and Hibbard, E. (1979): An SEM and TEM study of suppression of eye development in eyeless mutant axolotl. *Anat. Embryol*. 156:29–35.

Vandervorst, P., and Ghysen, A. (1980): Genetic control of sensory connections in *Drosophila*. *Nature* 286:65–67.

Van Deusen, E. (1973): Experimental studies on a mutant gene (e) preventing the differentiation of eye and normal hypothalamus primordia in the axolotl. *Dev. Biol*. 34:135–158.

Van Houten, J. (1977): A mutant of *Paramecium* defective in chemo-taxis. *Science* 198:746–748.

Van Houten, J. (1978): Two mechanisms of chemotaxis in *Paramecium*. *J. Comp. Physiol.* 127A:167–174.

Van Houten, J. (1979): Membrane potential changes during chemo-kinesis in *Paramecium*. *Science* 204:1100–1103.

Van Houten, J., Chang, S. Y., and Kung, C. (1977): Genetic analyses of 'paranoiac' mutants of *Paramecium tetraurelia*. *Genetics* 86: 113–120.

Van Houten, J., Hansma, H., and Kung, C. (1975): Two quantitative assays for chemotaxis in *Paramecium*. *J. Comp. Physiol.* 104: 211–223.

Venard, R. (1980): Attractants in the courtship behavior of *Drosophila melanogaster*. In *Development and Neurobiology of Drosophila*. Siddiqi, O., et al., eds. New York: Plenum Press, pp. 457–465.

Venard, R., and Jallon, J.-M. (1980): Evidence for an aphrodisiac pheromone of female *Drosophila*. *Experientia* 36:211–213.

Vogt, W. (1925): Gestaltungsanalyse am Amphibienkeim mit örtlicher Vitalfarbung. I. Teil. Methodik und Wirkungsweise der örtlichen Vitalfarbung mit Agar als Farbtrager. *Wilhelm Roux' Arch. Entwicklungsmech. Org.* 106:542–610.

Wadepuhl, M., and Huber, F. (1979): Elicitation of singing and court-ship movements by electrical stimulation of the brain of the grass-hopper. *Naturwissenschaften* 66:320–322.

Walton, J. N., and Gardner-Medwin, D. (1974): Progressive muscular dystrophy and the myotonic disorders. In *Disorders of Voluntary Muscle,* 3rd ed. Walton, J. N., Edinburgh: Churchill Livingstone, pp. 561–613.

Wang, E. A., and Koshland, D. E., Jr. (1980): Receptor proteins in the bacterial sensing system. *Proc. Nat. Acad. Sci.* 77:7157–7161.

Ward, S. (1973): Chemotaxis by the nematode *Caenorhabditis elegans:* Identification of attractants and analysis of the response by use of mutants. *Proc. Nat. Acad. Sci.* 70:817–821.

Ward, S. (1976): The use of mutants to analyze the sensory nervous system of *Caenorhabditis elegans*. In *The Organization of Nematodes*. Croll, N. A., ed. New York: Academic Press, pp. 365–382.

Ward, S. (1977): Invertebrate neurogenetics. *Ann. Rev. Genet.* 11: 415–450.

Ward, S. (1978): Nematode chemotaxis and chemoreceptors. In *Taxis and Behaviour.* (Receptors and Recognition, Series B, Vol. 5.) Hazelbauer, A. L., ed. London: Chapman and Hall, pp. 141-160.

Warrick, H. M., Taylor, B. L., and Koshland, D. E. (1977): Chemotactic mechanism of *Salmonella typhimurium:* Preliminary mapping and characterization of mutants. *J. Bacteriol.* 130:223-231.

Waterston, R. H. (1981): A second informational suppressor, *sup-7 X,* in *Caenorhabditis elegans. Genetics* 97:307-325.

Waterston, R. H., and Brenner, S. (1978): A suppressor mutation in the nematode acting on specific alleles of many genes. *Nature* 275: 715-719.

Waterston, R. H., Fishpool, R. M., and Brenner, S. (1977): Mutants affecting paramyosin in *Caenorhabditis elegans. J. Mol. Biol.* 117: 679-697.

Waterston, R. H., Thomson, J. N., and Brenner, S. (1980): Mutants with altered muscle structure in *Caenorhabditis elegans. Dev. Biol.* 77: 271-302.

Wecker, L., Laskowski, M. B., and Dettbarn, W.-D. (1978): Neuromuscular dysfunction induced by acetylcholinesterase inhibition. *Fed. Proc.* 37:2818-2822.

Weimar, W. R., and Sidman, R. L. (1978): Neuropathology of "vibrator"—a neurological mutation of the mouse. *Soc. Neurosci. Abstr.* 4:401.

Weisblat, D. A., Sawyer, R. T., and Stent, G. S. (1978): Cell lineage analysis by intracellular injection of a tracer enzyme. *Science* 202: 1295-1298.

Weiss, P. A. (1955): Nervous system (neurogenesis). In *Analysis of Development.* Willier, B. H., Weiss, P. A., and Hamburger, V., eds. Philadelphia: W. B. Saunders, pp. 346-401.

Weiss, P. A. (1971): Nervous system (neurogenesis). In *Analysis of Development.* Willier, B. H., Weiss, P.A., and Hamburger, V., eds. New York: Hafner, pp. 346-401.

Welshons, W. J. (1965): Analysis of a gene in *Drosophila. Science* 150:1122-1129.

West, G. J., and Catterall, W. A. (1979): Variant neuroblastoma clones with missing or altered sodium channels. *Proc. Nat. Acad. Sci.* 76:4136-4140.

Wheeler, G. L., and Bitensky, M. W. (1977): A light-activated GTPase in vertebrate photoreceptors: Regulation of light-activated cycle GMP phosphodiesterase. *Proc. Nat. Acad. Sci.* 74:4238-4242.

White, J. G., Albertson, D. G., and Anness, M. A. R. (1978): Connectivity changes in a class of motoneurone during the development of a nematode. *Nature* 271:764-766.

White, J. G., Southgate, E., Thomson, J. N., and Brenner, S. (1976): The structure of the ventral nerve cord of *Caenorhabditis elegans*. *Phil. Trans. Roy. Soc. B* 275:327-348.

White, K. (1980): Defective neural development in *Drosophila melanogaster* embryos deficient for the tip to the *X* chromosome. *Dev. Biol.* 80:332-344.

White, K., and Kankel, D. R. (1978): Patterns of cell division and cell movement in the formation of the imaginal nervous system in *Drosophila melanogaster*. *Dev. Biol.* 65:296-321.

White, R. H., and Lord, E. (1975): Diminution and enlargement of the mosquito rhabdom in light and darkness. *J. Gen. Physiol.* 65: 583-598.

Wilcox, M. J. (1980): Ionic mechanism of the receptor potential in the photoreceptors of wild-type and mutant *Drosophila*. Ph.D. Thesis, Rockefeller University, New York, New York.

Williamson, R., Kaplan, W. D., and Dagan, D. (1974): A fly's leap from paralysis. *Nature* 252:224-226.

Wilson, B. W., Randall, W. R., Patterson, G. T., and Entrikin, R. K. (1979): Major physiologic and histochemical characteristics of inherited dystrophy of the chicken. *Ann. N. Y. Acad. Sci.* 317: 224-246.

Wolf, R., Gebhardt, B., Grademann, R., and Heisenberg, M. (1980): Polarization sensitivity of coarse control in *Drosophila melanogaster*. *J. Comp. Physiol.* 139:177-191.

Worden, F. G., Childs, B., Matthysse, S., and Gershon, E. S. (1976): Frontiers of psychiatric genetics. *Neurosci. Res. Prog. Bull.* 14:1-107.

Wright, T. R. F. (1970): The genetics of embryogenesis in *Drosophila*. *Adv. Genet.* 15:261-295.

Wright, T. R. F. (1977): The genetics of dopa decarboxylase and $\alpha$ -methyl dopa sensitivity in *Drosophila melanogaster*. *Am. Zool.* 17: 707-721.

Wright, T. R. F., and Bentley, K. W. (1979): The effects of a temperature-sensitive dopa decarboxylase deficient allele on female fertility in *Drosophila*. *Genetics* 91:s139-s140.

Wright, T. R. F., Bewley, G. C., Sherald, A. F. (1976): The genetics of dopa decarboxylase in *Drosophila melanogaster*. II. Isolation and characterization of dopa-decarboxylase-deficient mutants and their relationship to the α-methyl-dopa-hypersensitive mutants. *Genetics* 84:287–310.

Wu, C. F., and Ganetsky, B. (1980): Genetic alteration of nerve membrane excitability in temperature-sensitive paralytic mutants of *Drosophila melanogaster*. *Nature* 286:814–816.

Wu, C. F., Ganetsky, B., Jan, L. Y., Jan, Y. N., and Benzer, S. (1978): A *Drosophila* mutant with temperature-sensitive block in nerve conduction. *Proc. Nat. Acad. Sci.* 75:4047–4051.

Wu, C. F., and Pak, W. L. (1975): Quantal basis of photoreceptor spectral sensitivity of *Drosophila melanogaster*. *J. Gen. Physiol.* 66: 149–168.

Yeh, J. Z., Oxford, G. S., Wu, C. H., and Narahashi, T. (1976): Dynamics of aminopyridine block of potassium channels in squid axon membrane. *J. Gen. Physiol.* 68:519–535.

Yoon, C. H. (1969): Disturbances in developmental pathways leading to a neurological disorder of genetic origin, "leaner," in mice. *Dev. Biol.* 20:158–181.

Young, M. W., and Judd, B. H. (1978): Nonessential sequences, genes, and the polytene chromosome bands of *Drosophila melanogaster*. *Genetics* 88:723–742.

Zettler, F., and Järvilehto, M. (1972): Lateral inhibition in an insect eye. *J. Vergl. Physiol.* 76:233–244.

## ADDENDA

Gill, K. S. (1963): A mutation causing abnormal courtship and mating behavior in *Drosophila melanogaster*. *Amer. Zool.* 3:507.

Hunt, R. K., and Ide, C. F. (1977): Radial propagation of positional signals for retino-tectal patterns in *Xenopus*. *Biol. Bull.* 153:430–431.

# NAME INDEX

# SUBJECT INDEX

Abdomen, and reproductive
  behavior, 108
*Ace* mutant, *Drosophila,* 54, 83,
  196–200
  mosaics, 56–60, 129–130
  optomotor behavior, 85–87
*ace* mutant, nematode, 62
Acetylcholine. *See* ACh
Acetylcholinesterase. *See* AChE;
  *Ace* mutant; *ace* mutant
ACh
  metabolism, 52–55
  in nematodes, 61–64
  receptor, 55–56, 63
AChE, 52–55, 121, 196–200. *See
  also Ace* mutant; *ace* mutant
  gynandromorphs, 107
  and mosaic mapping, 129–130
  in nematode, 62
Acid phosphatase, 121–123
Action potential
  and facilitation, 50
  miniature, 50–51
  mutants, 41–49
  *Paramecium,* 33–34
Adaptation, 8
Adenine, 119
Affective disorders, 117–118
Aggression, 114–115
Albino mutants, 177–180
Allophenics, 152
*Ambystoma,* 113–114, 202
*A. mexicanum,* 49
*A. punctatum,* 150, 151
*A. tigrinum,* 150, 151
Aminopyridine, 45–47, 199–200
amnesiac mutant, 95, 112
Amphibians
  CNS mosaics, 157–159
  development, 148–153, 200–202
  neurospecificity, 194–197
  visual system, 148–153

Aneuploids, 52
Antennapedia (*Antp*) mutant, 165,
  170–171
anterobithorax (*abx*) mutant, 168
Antidepressants, 117
Anurans, 150
*Aplysia,* 91
Apomorphine, 119
aristapedia (*ss^a*) mutant, 165,
  170–171
*Ascaris,* 62, 64
Aspartate, 10
Associative conditioning, 93
ATP, 35
Attractants, 10, 14–17, 18–19, 66–69
Auditory transduction, 32
Avoiding reaction, 18–21, 37, 39–40
Axolotl, 49, 202
  mosaic studies, 136
  mutants, 113–114, 148–149,
    159–160, 187
Axons, 162–163

Bacterial chemotaxis, 8–17, 66–69
bang-sensitive mutant, 50
Barium resistance, 36
Basal ganglia, 119
Batrachotoxin-resistant cells, 43
Behavior. *See also* Learning;
    Memory, and learning
  bacterial, 8–11, 66–69
  chemosensory, 18–24, 108–111
  general, 69–75
  higher, neurochemistry of, 114–120
  optomotor, 78–79
  reproductive, 98–114
Behavioral mutants, 1–3
  *Drosophila,* 71–75
  and enzymes, 62–63
  in mosaics, 127
  paramecia, 34
Behavioral neurogenetics, 66–120

277